T0314461

BUSINESS MEASUREMENTS
FOR
SAFETY
PERFORMANCE

BUSINESS
MEASUREMENTS
FOR
SAFETY
PERFORMANCE

Daniel Patrick O'Brien, CSP

LEWIS PUBLISHERS

Boca Raton London New York Washington, D.C.

Library of Congress Cataloging-in-Publication Data
Catalog record is available from the Library of Congress

This book contains information obtained from authentic and highly regarded sources. Reprinted material is quoted with permission, and sources are indicated. A wide variety of references are listed. Reasonable efforts have been made to publish reliable data and information, but the author and the publisher cannot assume responsibility for the validity of all materials or for the consequences of their use.

Neither this book nor any part may be reproduced or transmitted in any form or by any means, electronic or mechanical, including photocopying, microfilming, and recording, or by any information storage or retrieval system, without prior permission in writing from the publisher.

The consent of CRC Press LLC does not extend to copying for general distribution, for promotion, for creating new works, or for resale. Specific permission must be obtained in writing from CRC Press LLC for such copying.

Direct all inquiries to CRC Press LLC, 2000 N.W. Corporate Blvd., Boca Raton, Florida 33431.

Trademark Notice: Product or corporate names may be trademarks or registered trademarks, and are used only for identification and explanation, without intent to infringe.

Visit the CRC Press Web site at www.crcpress.com

© 2000 by CRC Press LLC
Lewis Publishers is an imprint of CRC Press LLC

No claim to original U.S. Government works
International Standard Book Number 1-56670-408-1
3 4 5 6 7 8 9 0

Dedication

This book is in memory of Dad, who taught me about integrity (Titus 2:7-8).

In appreciation of Gloria, who demonstrates day by day what a real Proverbs 31 woman is all about (Proverbs 31:10-31).

And as encouragement to Tyson, James, and Madison, to show them that all things are possible to those who believe (Matthew 21:21).

Contents

About the Author

Dan O'Brien is a Certified Safety Professional, and is currently the safety and health manager for Engineered Carbons, Inc. in Borger, Texas. He is a professional member and past president of ASSE's Panhandle Chapter. He holds a BS in Industrial Education, and an MS in Industrial Technology, both from West Texas State University. O'Brien is currently serving as Secretary for the North American Product Safety and Regulatory Committee for the International Carbon Black Association, where he also serves on the Industrial Hygiene, NIOSH, TLV, and Hazcom Subcommittees. He is an Adjunct Professor at West Texas A&M University.

Foreword

Let's hear the number—how many widgets did you make? How many pounds of product did you produce? It's performance review time for your production department and the only number management cares about is the number of total widgets out the door—and the total pounds produced! Production department performance is evaluated solely on this number. There is no reference made or importance given to the amount of raw material that was used to make the corresponding widgets. On-time shipments and correct or offspec production are of little or no consequence. Whether production equipment was maintained or even ruined has no bearing on the performance review of this production department. Personnel issues such as employee safety, employee turnover, absenteeism, and overtime are not considered. Widgets out the door/pounds produced—that's all! *Not the way production is measured in your company?* Maybe the sales department will measure closer to your company's performance criteria. Our sales department's performance is measured solely according to the number of widgets/pounds of product sold. Similarly, no consideration is given to the profit margin obtained, whether sales matched production, or if regular customers were supplied. The number of sales calls made, customer service, and delivery schedules committed play no part in how this sales department is measured. Most management would be appalled at the mere mention of such a closed view of performance measurement.

Every business sector employs methods of measurement very different from those in the two examples just given. Whether we consider production, sales, accounting, research, marketing, engineering, etc., a multitude of metrics are used to evaluate the performance of a particular business sector; *that is, except safety!* In his book *Axioms of Industrial Safety*, Heinrich pointed out decades ago: "The most valued methods in accident prevention are analogous with the methods for the control of quality, cost and quality of production."[1] In today's business culture, one single number, OSHA recordable, typically measures safety. Even stretching, we would only include lost time accidents or maybe medical cases. Is this really how we want to run our businesses? I contend it is not! After all, safety affects the bottom line of a company's profitability just as any other business sector does. If you don't believe it, examine your current loss run statement from your insurance carrier. Reviewing the cost of a strained back, an employee on long-term disability, lost production, or damaged equipment from accidents will quickly convince you that safety is most definitely a business sector. The question then is: how do we quantifiably measure safety performance in ways similar to the ones in which we measure production performance, sales performance, or any other business sector? This text gives a simple, specific, and quantitative method to measure the safety business sector. Safety performance, presented in the same terms as other business sectors, is measured. The same "most valued methods" utilized by other business sectors can now be adopted to improve overall safety performance.

1. Heinrich,H.W.: Industrial Accident Prevention, McGraw-Hill, New York, 1st ed. 1931, 2nd ed. 1959

Acknowledgments

This book was inspired by many of the country's best-known safety "consultants." While that may be their official career title, in most cases "teacher" would be much more appropriate. Masters of the safety profession, like E. Scott Geller and Dan Petersen, have encouraged me to think of safety in ways not typically considered. Their constant challenge to view safety from a different perspective, reexamine current beliefs, and embrace new methods and practices has been critical in the development of these safety performance measurement criteria. Even their open disagreement on specifics has encouraged a willingness to look beyond "how we've always done it." Dave Johnson has been instrumental in helping me take notes from "down in the trenches" and transform them into usable information from which all can benefit.

I must also acknowledge my wife Gloria. Without her behind the scenes, and her hard work and diligent efforts to put all the loose ends together, this project would not have been possible.

1

The History of Performance Measurement

By whatever name you understand this new frontier, it is the most far-reaching revolution in history. It is bigger than any spiritual, sexual, industrial, or business revolution in the history of mankind.

Daniel Patrick O'Brien

Performance measurement is not some revelation of 1990s management. Performance measurement has been around as long as man has existed. Starting with Adam and Eve, man's performance has been measured against given criteria. God said, "You shall not eat it (the fruit)."[1] Whether or not they ate the fruit was the "performance" being measured. While this may seem simplistic, it gives a sense of just how basic our need for performance measurement is. In the most basic forms of performance measurement, man has always been measured and always will be measured. Without measurement there is no way of knowing how good or bad an individual or process is doing. Constant improvement and eventual zero-defect production are almost impossible without a good performance measurement system in place.

Measurement is the first step that leads to control and eventually to improvement. If you can't measure something, you can't understand it. If you can't understand it, you can't control it. If you can't control it, you can't improve it.[2]

H. James Harrington

The bible is full of references to "performance measurement." In the familiar parable of the talents, the three men were held accountable for the return on the money for which they were given responsibility. Their "performance" was not arbitrarily rewarded. The "top performer" was rewarded with the top compensation—promoted in today's terms. The worst performer was rewarded (punished) accordingly—demoted or fired in today's terms.[3] One final example from the Bible: in about 605 BC, Daniel's consistent "performance" as leader, manager, and faithful employee landed him the most prestigious and powerful job in all of Babylon.[4] Obviously, his performance had been closely measured before he ascended to such a lofty position.

Try to imagine the Great Pyramids of Egypt being built with no measure of performance. These historical builders surely had just as much need for performance measurement as industry does today. The pyramid builders had time and budget commitments to keep, manpower "head counts" and production goals to meet. You can be certain the paymaster for the Great Pyramids did not continue to pay wages for unproductive workers.

Throughout the Middle Ages a system of master and apprentice was utilized in all trades. Trades were the forerunners of modern industry and labor. This system was not merely one of training, but a system of "measuring" one's skill, talent, knowledge, and abilities. An apprentice could not advance to the level of master until his "performance measurement" was of a suitable level and quality.

Prior to the twentieth century, the value of a slave was determined, among other things, on his or her performance. Slaves with reputations for being hard workers and good producers, and who were fit and able- bodied, would bring premiums over slaves who did not have the same ability to attain high levels of "performance."

Before the start of the Industrial Revolution, performance measurement was largely centered on the capabilities of a single person. While the performance of a group might still have been measured, it was the ability of the individual that was measured. Until the 1700s, production methods were labor-intensive, with work being done by hand in cottages. Three developments were to change this way of life:

1. In England, inventors developed the Spinning Jenny in 1764.
2. The power loom was perfected in 1784.
3. In America, Eli Whitney developed the cotton gin in 1792.

These and other innovations ushered in what would later be known as the Industrial Revolution.[5]

Performance measurement as we know it today really began with the advent of the Industrial Revolution. The Industrial Revolution, coupled with mass production techniques, brought about a need for statistical performance measurements. New machines allowed faster and more uniform production of a wide range of products and processes. This made comparisons between employees, equipment lines, and process methods possible, even necessary, in order to ensure production was at the highest level possible. While previous performance measurements focused predominately on output and individual performance, more sophisticated production capabilities promoted the need for more sophisticated means of measurement.

Industrial capabilities, scientific advancements, and computer technology were all rocketing to new and previously unknown frontiers starting in the early 1900s. Gaining unbelievable momentum in the 1970s, 1980s, and 1990s, the Industrial Revolution gave way to what some refer to as the "Business Revolution of the 1990s" or the "Information Age." The Information Age has given all levels of an organization access to the measurement results of individuals, groups, processes, business units—even entire industries—that were previously not possible. What electricity did for the light bulb, the Information Age is doing for performance measurement. The uses and benefits of performance measurement go hand-in-hand with the benefits of the Information Age as a whole. With current process controls, automated tracking systems, and continual and effortless data-logging on every front, performance measurement has firmly taken its place in the growth of mankind and its advancements.

This new level of data availability brought about many measuring tools — statistical process control charts (SPC), flow diagrams, Gant charts, Pareto charts, and fault-tree analyses, just to mention a few in wide use today. Performance appraisals, Management by Objective, Behavior Modification, and Competency-Based Performance Management are additional tools used to measure performance in this new and faster world of information availability. The capabilities and availability to measure progress truly are revolutionary. By whatever name you understand this new frontier, it is the most far-reaching revolution in history. It is bigger than any spiritual, sexual, industrial, or business revolution in the history of mankind. These new frontiers that were considered science fiction just a few years ago are real-time, real-life happenings today, and they must be measured in order to understand and control them effectively.

We see through this brief summary of the evolution of performance measurement that it is now a viable, everyday tool in an organization. Unfortunately, it is not uniformly and systematically used throughout industry. Performance measurement is so diverse that even similar organizations may use drastically different performance-measurement tools. What works for one individual, group, or organization may not work for another; what is too advanced for one group may be too simplistic for another. You see this diversity in the different ways companies measure their processes, production, and employees. It is still commonplace to see this diversity of performance-measurement techniques within the same organization. Production, for example, might be measured by the number of widgets produced, while sales are measured by dollar value of sales. Within the same department you might see the number of shipments made each day compared to the number of on-time shipments. While there may be a predominating measure in each of these examples, most likely there are multitudes of other measures by which each "business unit" is measured. Typical business-unit measurement often tends to be more of a shotgun approach than a rifle approach. A typical business unit might be measured by sales volume, on-time shipments, off-spec products, damaged equipment, personnel problems, net profit, and many, many other metrics. While virtually all other business sectors have a multitude of metrics by which they are judged, safety performance is most often measured by one metric: the OSHA recordable incidence rate. In the next chapter we will review how safety performance has evolved.

NOTES AND REFERENCES

1. Scripture taken from the *New King James Version*. Copyright ©1982 by Thomas Nelson, Inc. Used by permission. All rights reserved. Genesis 3:3.
2. Kaydos, Will J., *Operational Performance Measurement; Increasing Total Production*, CRC Press, Boca Raton, FL, 1999, p. 3.
3. Scripture taken from the *New King James Version*. Copyright ©1982 by Thomas Nelson, Inc. Used by permission. All rights reserved. Matthew 25:14–30.
4. Scripture taken from the *New King James Version*. Copyright ©1982 by Thomas Nelson, Inc. Used by permission. All rights reserved. Daniel 2:46, 6:26.
5. National Safety Council, *Accident Prevention Manual For Business and Industry, Administration and Programs*, 11th edition, 1997, p. 4.

2

The Beginning of Safety Performance Measurement

Companies found they could no longer ignore the social, moral, ethical, and financial ramifications of not providing a safe workplace for all employees.

Daniel Patrick O'Brien

As we learned in the last chapter, performance measurement has been around since the beginning of time. While, in the strictest sense, safety-performance measurement has also been around that long, true, specific measurement in the safety area is a relatively new endeavor. Early attempts to measure safety performance were little more than a tally of how many fatalities an organization had suffered. As the Industrial Revolution advanced and more and more mechanical equipment came into the workplace, safety-performance measurement began to include amputations and other serious workplace accidents, but little or no attention was given to eliminating hazards or monitoring safety-related behaviors and activities.

When we think of safety measurement in industry today, we immediately think of measuring everything that could possibly influence the injury or illness of workers, as well as factors that affect the equipment, environment, and surrounding communities. While noble and certainly more appropriate, this is far from the original concept of safety-performance measurement, even before it was called "safety-performance measurement." Safety measurement first came to the surface as something worthy of measurement only because of the severe consequences of not measuring. The only metrics considered worthy of measurement were fatalities, serious injuries, or those involving major equipment loss. One must remember that the industrial landscape was much different than the industrial landscape we know today. The industrial setting was filled with an entirely different culture, mindset, experience, and knowledge than is prevalent today.

As true safety-performance measurement first came on the scene there was a workforce comprised predominantly of women and children. The 1900 census showed 1,750,178 working children between 10 and 15 years of age.[1] Their work day was basically twice as long as a normal work day today. Women and children would commonly work from before sunrise until late at night, most often in areas that had poor lighting, bad ventilation, unsafe noise levels, and many other work conditions that are considered unacceptable in today's workplace. There was little or no safeguarding of machinery or any other considerations for the basic safety of the worker. Amputations, maiming, and death were commonplace, if not the norm. The rapid expansion of mechanical equipment made production and processing achievable at previously impossible rates. There was a focus of "production first and foremost," which continued without challenge until late in the 1800s. Only then did the welfare of the worker begin to be acknowledged and specifically measured. Even then, sad as it is, the well-being of the worker was not the primary driving force in changing the "production above all else," culture. The driving force most often was the employer's inability to replace maimed or killed workers. Something had to be done.

In contrast to the modern-day work environment, Figures 1 through 3 show examples of working conditions during this time period. Nevertheless, in the late 1800s and early 1900s, changes slowly began to be made in the workplace to provide safer conditions for workers. Though changes for the benefit of workers were beginning to be made, the changes were quite basic and rudimentary compared to modern-day safety and health

FIGURE 2.1 Differs in Cherryville Manufacturing Company, 1908. (Photograph courtesy of the National Archives)

FIGURE 2.2 Boys, 7 and 12 years old, in Roanoke, Virginia, 1911. (Photograph courtesy of the National Archives)

FIGURE 2.3 A 12-year-old girl spinner in a cotton mill in North Pownal, Vermont, in 1910. (Photograph courtesy of the National Archives)

programs. Some of these changes were due to new laws being enacted, others were the result of the first worker compensation laws in the United States. Unions were also forming, which brought hazardous conditions to a new light. Still more changes were the result of companies beginning to understand the bottom-line cost consequences of accidents and injuries in the workplace. All of these factors helped to increase the awareness of safety in the eyes of workers, companies, communities, and the world.

In 1867, Massachusetts began using factory inspectors, some of the first in the nation, to point out unsafe work conditions. This was really the first major move toward a "governmental-compliance" mentality. In 1877, Massachusetts passed the Employer's Liability Law, the first U.S. law that made employers responsible for implementing safeguards on machinery. Employers were forced to take a cursory step toward a safer workplace. It was at that time in history when performance measurement began to specifically look at safety.

In 1908 President Theodore Roosevelt stated:

> The number of accidents which result in the death or crippling of wage earners is simply appalling. In a very few years it runs up a total far in excess of the aggregate of the dead in any major war.[2]

The first bill for workers' compensation (the Wainwright Act) was passed in New York in 1910, but it was declared unconstitutional by the New York Court of Appeals. The court held that the law violated both the federal and New York State constitutions, "because it took property from the employer and gave it to the employee without due process of law." On the same day the 1910 act was declared unconstitutional, March 25, 1911, a devastating fire in New York City's Triangle Clothing Factory killed 146 workers. This disaster, called the Triangle Factory Fire, outraged the public and spurred a demand for factory legislation and health and safety reform. After an amendment to the state constitution was approved in 1913 in the general election, a compulsory Workmen's Compensation Act finally became effective in 1914.[3] Workers' compensation legislation provided the financial atmosphere for industrial safety to finally be recognized as a worthy endeavor for companies. Workers' compensation, in effect, states that regardless of fault, the injured employee will be compensated for injuries that occur on the job.[4]

In 1913 the National Safety Council was formed. One of its first missions was to survey the state, federal, and municipal regulations regarding safety. By 1918 it was clear to all that the safety of the national workplaces was a shambles. The survey helped to combine the efforts of national and state governments, engineering societies, insurance companies, unions, and other groups to create national regulations and standards. These standards are the forerunners of many of the laws and standards in existence today.

As state laws declared protection for employees, higher courts typically overruled the lower court decisions. While this battle continued back and forth, industry slowly began to realize the benefits and profitability of stopping injuries before they occurred. Gradually, the laws began to sway in favor of protecting the worker. Various factors contributed to this: concern for the employee; a quest to improve profitability; and mandates from federal, state, and local regulations. Insurance companies began to take active roles in pushing employee safety to new frontiers. Child labor laws also impacted on safety in the workplace. Laws protecting the worker and holding the employer responsible for injuries suffered by the employee were being enacted across the nation. Driven by financial ramifications, new regulations, public sentiment, and worker demands, companies began initiating practices that would dramatically improve the workplace safety of employees. Safety performance measurement was finally beginning to take shape.

It was in the early 1900s that the "Three E's" of safety came into play—Engineering, Education, and Enforcement. Prior to this time there were no laws to abide by, no devotion to engineering controls, and certainly no perceived benefit to training. Training was limited to the information necessary to perform the task and not much else. Performance measurement in safety was beginning to make a significant contribution to the overall working conditions of workers across the nation and around the world.

In 1931, H. W. Heinrich published *Industrial Accident Prevention.*[5] In this text Heinrich presented his original "Axioms of Industrial Safety." The axioms are worth reprinting to demonstrate just how quickly industry had moved from focusing little or no attention on the welfare and safety of its employees, to beginning to devote significant amounts of energy toward preventing accidents. While research has since shown some of these axioms do not reflect the most recent trends in safety and/or management thinking, they clearly show a specific and concentrated advancement in safety performance measurement efforts. Current axioms of safety will be discussed in a later chapter.

Heinrich's axioms helped to start a major effort in the 1930s and 1940s to remove hazardous conditions from the workplace

Axioms of Industrial Safety

1. The occurrence of an injury invariably results from a completed sequence of factors — the last one of these being the accident itself. The accident in turn is invariably caused or permitted directly by the unsafe act of person and/or a mechanical or physical hazard.

2. The unsafe acts of persons are responsible for a majority of accidents.

3. The person who suffers a disabling injury caused by an unsafe act, in the average case has had over 300 narrow escapes from serious injury as a result of committing the very same unsafe act. Likewise, persons are exposed to mechanical hazards hundreds of times before they suffer injury.

4. The severity of an injury is largely fortuitous—the occurrence of the accident that results in injury is largely preventable.

5. The four basic motives or reasons for the occurrence of unsafe acts provide a guide to the selection of appropriate corrective measures.

6. Four basic methods are available for preventing accidents— engineering revision, persuasion and appeal, personnel adjustment, and discipline.

7. Methods of most value in accident prevention are analogous with the methods required for the control of the quality, cost, and quantity of production.

8. Management has the best opportunity and ability to initiate the work of prevention; therefore it should assume the responsibility.

9. The supervisor or foreman is the key man in industrial accident prevention. His application of the art of supervision to the control of worker performance is the factor of greatest influence in successful accident prevention. It can be expressed and taught as a simple four-step formula.

10. The humanitarian incentive for preventing accidental injury is supplemented by two powerful economic factors: (1) the safe establishment is efficient productively and the unsafe establishment is inefficient; and (2) the direct employer cost of industrial injuries for compensation claims and for medical treatment is but one-fifth of the total cost which the employer must pay.

and began an effort to train workers on safe ways of performing their work. Heinrich's axioms were also among the first publications to place the responsibility of the safety of the worker on supervision and management. With the new emphasis on protecting workers came a new effort to control occupationally related illnesses and diseases.

While recognition and control of occupational illnesses and occupational-related diseases was not anything new, the focus and emphasis were definitely novel . Occupational diseases have been recognized since the beginning of civilization. Hypocrites wrote in 500 BC that many miners had breathing difficulties, and by 100 BC respirators were in use by miners to prevent the inhalation of dust. In 1700 Ramazzini wrote a comprehensive book on occupational medicine in which he identified specific diseases related to certain occupations. Until the twentieth century, physicians were the primary group interested in occupational diseases. An interest in occupational diseases was thrust on the safety professional when they became compensible in the early 1930s.[6] In 1939, the American Industrial Hygiene Association was established to promote the recognition, evaluation, and control of environmental stresses arising in or from the workplace. Safety-performance measurement was taking serious shape and form.

By the 1940s and 1950s, safety-performance measurement was developing alongside engineering disciplines, health care, management, and other developing disciplines. Safety was now being viewed as an integral part of business parameters. Safety professionals were broadening the scope of what was traditionally considered to be safety. Safety professionals found themselves dealing with industrial hygiene issues, management, financial participation, human resource issues, engineering and design issues, risk management, and a whole host of other activities and issues that previously were not considered in the scope of safety performance. Safety had become recognized as not only a financially beneficial effort, but also essential from a multitude of different perspectives. Companies found they could no longer ignore the social, moral, ethical, and financial ramifications of providing a safe workplace for all employees. This meant measuring safety performance in specific terms and using varied methods. Safety professionals were beginning to see that employee safety off the job affected on-the-job safety performance. The National Safety Council's *Accident Facts 1998 Edition* reported that while off-the-job injuries were about 5,800,000, workplace injuries were about 3,800,000. Deaths showed a similar ratio of off-the-job deaths versus on-the-job deaths—38,200 to 5,100, respectively.[7] These are current statistics, but the same trends were first noticed in the 1950s and 1960s. These trends further promoted the safety professional into new areas of expertise, time commitment, and new considerations in safety performance measurements.

In 1970, the U.S. government further heightened the focus on workplace safety and health. This was the year that the Williams-Steiger Occupational Safety and Health Act was passed, creating the Occupational Safety and Health Administration, commonly known as OSHA. This was the first time that the United States had a comprehensive national safety law. OSHA focused primarily on physical hazard abatement. This forced the safety professional to concentrate on the removal of physical hazards and on documentation. OSHA regulations (standards) forced compliance on seemingly endless fronts. Often, the safety professional was divided in two. Compliance issues were not always the same as controlling losses but, nevertheless, OSHA forced the national safety and health initiative to new levels of safety-performance measurement.

Worldwide, through the 1970s, 1980s, and 1990s, countries were passing national safety and health initiatives. Some dealt solely with safety and health issues, others placed more emphasis on environmental issues, and still others leaned more toward industrial-hygiene standards. Regardless of their specific focus, there was a worldwide effort to push safety and health issues to new levels of safety-performance measurement.there was a worldwide effort to push safety and health issues to new levels of safety-performance measurement.

Major Safety and Health Legislation Worldwide

United Kingdom	Health Safety and Work Act	1974
United States	Clean Air Act	1990
Australia	Victoria Occupational Health and Safety Act	1985
Canada	Worker's Compensation Act	1990
Canada	Ontario Occupational Health and Safety Act	1978
United Kingdom	Health and Safety at Work Act	1974
United States	Superfund Amendments and Reauthorization Act	1986
Canada	Workplace Hazardous Management Information System	1985
United States	Resource Conservation and Recovery Act	1976
United States	Comprehensive Environmental Response Compensation and Liability Act	1980
United Kingdom	Framework Directive	1989
United Kingdom	International Standards Organization	1996

NOTES AND REFERENCES

1. National Safety Council, *Accident Prevention Manual for Business and Industry, Administration and Programs*, 11th edition, 1997, p. 6.
2. National Safety Council, *Accident Prevention Manual for Business and Industry, Administration and Programs*, 11th Edition, 1997, p. 7.

3. National Safety Council, *Accident Prevention Manual for Business and Industry, Administration and Programs*, 11th edition, 1997, p. 7.
4. Petersen, Dan, *Safety Management: A Human Approach*, 2nd edition, Aloray Inc., Goshen, NY, 1988, p. 3.
5. Heinrich, H.W., Petersen, Dan, Nestor, Ross, *Industrial Accident Prevention—A Safety Management Approach*, 5th edition, McGraw-Hill Book Company, New York, 1980, p. 21.
6. Petersen, Dan, *Safety Management: A Human Approach*, 2nd edition, Aloray Inc., Goshen, NY, 1988, p. 4.
7. The National Safety Council, *Accident Facts 1998 Edition*, p. 29.

3

Result-Oriented and Reactive Measurements

While OSHA Recordable Rates are helpful, they are similar to steering your car by using the rearview mirror to navigate where you're going. The mirror gives a good view of where you've been, but provides little assistance in steering toward new destinations.

Paraphrased from W. Edward Deming

One of the biggest obstacles to improving safety is safety measurement. This has always been the case, and continues today. Safety has not yet been integrated into the corporate culture of organizations. There are great campaigns and slogans: "Safety is Number- 1 at Our Company," "Safety First," and "Accidents Stop with You." These efforts are all commendable and give an appearance of attention and focus, but they do little to change the underlying culture of an organization. To further hinder the progress of safety, the measurements used to gauge safety performance are results-oriented and reactive. They are "trailing indicators," that is, they measure past efforts, loss events, problem areas, and past trends. They are totally dedicated to how things were, not how things are. Other trailing indicators might include epidemiology studies, worker compensation costs, regulatory citations, and cost of litigation.

Fatalities are naturally among the worst fears of an organization. They are certainly worth tracking, measuring, and attempting to avoid, but using them as a gauge of safety performance is simply not valid. A supervisor may work his or her entire lifetime, never do a single thing to raise the level of safety, and still never have a fatality in his or her department. Depending on fatalities as a gauge of safety performance would be like measuring the boiler operator by how many times he or she blows up the boiler.

As we begin to look in-depth at Business Measurements for Safety, we should spend some time discussing the different mindsets when it comes to performance measurement. Obviously, not all attempts to measure performance use the same techniques, methodologies, or systems. They are also not done from the same perspective; that is, different performance measurement techniques often come from vastly different schools of thinking. For example, measuring a football team's win–loss record is substantially different from measuring the team's ratio of third-down completions, as is measuring the number of training hours spent on kickoff returns. It is dangerous for us to consider all measurement tools and metrics together as if they were all equal. Each performance measurement is valid and useful to whatever degree of usefulness it was designed. Understanding this basic difference is important.

Some performance-measurement tools are single-purpose metrics, e.g., who won the Superbowl (or how many fatalities your company had). Other metrics are solely used for internal comparison. In our football example above, internal metrics might be used to determine which of the team's running backs should be used in short-yardage attempts close to the goal line (or which department conducted the most safety meetings). Still other metrics are used for external comparisons. To use our football example one more time, external comparisons might be very useful in determining whether the team's running back is performing at league standards of achievement (or which company leads its industry in accident reduction). While these examples may have no bearing on the caliber of the team's running back, they help us understand that performance measurement is dramatically different depending on *what is being measured* and *for what purpose it is being measured*. Let's look specifically at a fundamental area for performance measurement in the safety industry that is a results-oriented and reactive measurement: the OSHA recordable incidence rate.

OSHA RECORDABLE INCIDENCE RATE

In the safety industry, the most prevalent results-oriented measurement is the OSHA recordable incidence rate. By definition, the OSHA recordable incidence rate is computed by the simple formula:

$$\frac{\text{Number of OSHA recordable incidence} \times 200,000}{\text{Total number of man-hours worked}}$$

This simple formula is probably the most recognized safety metric in use in industry today. It quickly and easily provides a basic metric by which a plant, a company, or an industry can measure its safety performance. Examining it closely, the formula is quite simple. The constant 200,000 used in the formula is the estimated number of hours that a workforce of 100 workers works in a typical year—that is, 100 workers, times 50 work weeks per year, times 40 hours per week of work:

$$100 \times 50 \times 40 = 200,000$$

This calculation standardizes accident tracking, regardless of the industry and the number of employees, and overlooks the seriousness of any particular injury or illness. In other words, a company with several thousand workers making widgets can be compared with a company with only 20 employees involved in some other industry completely uninvolved with the widget business. The OSHA recordable incidence rate provides a computed number, the incidence rate, that, in short, shows how many workers out of 100 were injured or were diagnosed with an illness over a one-year period. An incidence rate of 6.0 would indicate that out of every 100 workers, six workers were injured seriously enough or had an illness that was serious enough to be logged onto the OSHA 200 report. The OSHA 200 formula is used to report and track injuries and illnesses. This safety performance metric is a simple and concise means of obtaining a measure of how well a company, plant, or industry is doing. The National Safety Council is one of many organizations that track

OSHA recordable rates for most industries. For example, the incidence rate for the manufacturing industry in 1996 was 10.6 compared with an incidence rate of only 2.4 in the finance, insurance, and real estate industries for the same year. Total incidence rates for all industries have fallen from 1973, when they were about 11, to a 1996 rate of just under 8.[1] The benefits and usefulness of this type of information are obvious. Its application is simple and straightforward and can be utilized by most anyone seeking to monitor "the big picture of safety."

THE OSHA RECORDABLE INCIDENCE RATE — RESULTS-ORIENTED

The OSHA recordable incidence rate measures what happened, not what is happening. This is not to say that measurements of this type are not valuable and useful, only that they provide little or no insight into real-time activities, behaviors, and cultures. In a "big-picture" view, certainly some sense of general direction and overall effectiveness can be derived about a safety program, but no operational or diagnostic measurement can be derived from them. "Results-oriented" is a general term used in industry to identify metrics of this nature. These indicators are ideal for a quick and broad snapshot of how the organization is doing in general terms. While the OSHA recordable rates are helpful, they are similar to steering your car by using the rearview mirror to navigate where you're going. The mirror gives a good view of where you've been, but provides little assistance in steering toward new destinations.

Dan Petersen gives this view in his book:

> *What measures should we use and why?* Are accident data useful for anything? It would save much time, effort and money to simply answer 'no', and focus on measures with meaning. However, it's not that simple. Many safety professionals do not agree—and most executives object to removal of these statistics. Despite these objections, it would still be best for practitioners to wean their companies from dependency on such worthless figures.[2]

Adding to the plight of using OSHA recordable incidence rates to measure a safety program is the fact that this metric is one of the easiest indicators or measurements to manipulate. At first sight this may not seem to be the case, but closer examination shows that the reporting of injuries and illnesses is easily skewed. Typically, this happens when employees or management drive reporting underground, inaccurately report injuries as "off-the-job injuries," and underreport injuries and illnesses via technicalities in OSHA's elaborate reporting system. Franklin E. Mirer, Ph.D., stated it this way:

> OSHA's recent embrace of injury rates as a performance measure echoes decades of bad management practice. Emphasizing lost workday injuries compounds the error by relying on the most easily manipulated outcome measure. By committing to an arbitrary target for national injury rate reduction, OSHA has doomed itself to failure.[3]

OSHA has made progress toward recognizing that the reporting of accidents hangs in a delicate balance. One criterion OSHA looks for in its Voluntary Protection Program (VPP) is whether any incentives or awards would inadvertently drive accident reporting underground. This only proves to be a small course correction on a journey going down the wrong road. True, the last thing a safety system needs to do is drive accident data to a point that the data is not reliable; more important is that the data most valuable to an organization be collected in the first place. Injury and illness incidence rates must be seen for what they are: a single snapshot of the safety culture, no more. These rates are still pictures of the safety culture.

We have metrics available that will measure more like a real-time video of the safety culture instead of a snapshot of how it was. Injury and illness incidence rate gives a results-oriented view of safety, a trailing indicator. There are other metrics that provide us with measurements of what is currently happening in the safety culture, not just what has happened in the past safety culture. We will examine this in-depth later. But first, let us look a little further into why injury incidence rates only show a tiny portion of the picture. Even then, the tiny picture may be a very misleading and limited performance indicator. Table 3.1 highlights some of the inadequacies in results-oriented and specific-indicator approaches to safety metrics. Obviously, these cases are exaggerated, but not unreasonable.

Summary of Company A

Company A has had a poor safety record in past years. The company has consistently had a higher insurance experience modifier because of this poor safety performance. Many of the more labor-intensive jobs are now performed by contract labor through the company's general contractor. This year, the contractor has had an OSHA recordable incidence rate of 18.9 and has suffered one fatality at the plant. Because of the fatality, OSHA recently cited the plant for 26 serious violations and 15 repeat violations, with fines totaling $2.6 million. The company's safety staff has been cut back to one person to cover six production sites. The safety person also handles human resources and environmental responsibilities at all six sites. No written safety policy exists. Company management has issued a "Commitment to Safety" statement highlighting how seriously they view accidents. No accidents have been *reported* since this "Commitment to Safety" statement has been issued.

Summary of Company B

Company B has had a consistent safety record for several years, with small decreases in OSHA recordable and medical cases each year. The company's insurance experience modifier has been below 1 for the last three years, and this year hit a low of 0.5. This has resulted in insurance cost savings of more than $100,000. All contractors who work in the company plants must participate in plant safety meetings. They must be audited by a team made up of safety personnel, management, and line employees, and must have an insurance experience modifier below 1 in order to work in any of the company locations. All contractor accidents are investigated as if they were company accidents. Company employees provide all safety training, with the plant safety manager acting as a facilitator. The company has a full range of safety policies, a complete auditing process for each, and quarterly compliance audits. The company is an OSHA VPP STAR plant. The company has had six accidents this year. Each accident was investigated within 24 hours by management-level personnel, safety personnel, and employees. Each shift is started with a five-minute "tailgate" safety meeting. Each employee is required to submit two safety observations during the month.

Company A and Company B: Equal or Drastically Different?

Classic results-oriented indicators for Company A and Company B show both companies are equal in safety performance. Both have an OSHA recordable incidence rate of 6.

TABLE 3.1   Side-by-Side Comparison of Two Companies

	Company A	Company B
OSHA recordables	6	6
Lost-time accidents	5	0
Restricted work cases	6	0
Experience modifier	1.9	0.5
Fatalities	1	0
Safety meetings held	6	300
Safety observations reported	0	200
Safety audits conducted	0	4
Contractor OSHA recordables	9	0
Contractor fatalities	1	0
Housekeeping audits	0	4
Training documentation	None	Complete
Management involvement	No	Yes
OSHA fines	$2.6MM	$0

Both companies have 100 employees.

That's where the similarities end. Insurance costs for Company A were approximately four times higher than those of Company B. Costs for lost time and production losses due to accidents are significant. OSHA fines and associated costs dramatically impact the bottom line. Employee involvement and, consequently, morale is as different as night and day. Company B's employees are constantly reminded of safety and are encouraged to seek out safety issues in a proactive way by submitting "safety observations." Company A has no system in place for controlling contractor accidents, whereas Company B makes contractors as accountable as the company employees must be. Company A management provides only lip service to safety versus Company B's active support and participation. Equal? Hardly!

It is apparent that limiting safety-performance measurements to only injury incidence rates is limited, at best. These corrective actions are typically "knee-jerk" responses by management. Figure 3.1[4] represents in graphical form this knee-jerk response typically caused by limiting safety-performance measurement to results-oriented and reactive measurements. This graph indicates increased management concern and activity when the frequency of injuries and illnesses increases. Once the system is "back in control," management typically decreases its effort, monetary outlay, and time commitment. This management surge and withdrawal is mirrored throughout the organization. As injuries and illness begin to increase again and head toward the upper limit, line management's emphasis on the safety program increases, starting the cycle over again. This may seem oversimplified, but it is the mindset of a large portion of management in today's industrial environment.

If an organization is only reactive, it cannot continuously improve. The very orientation of how it responds to market changes, unexpected losses, and even plans for the future are based on reacting to the changing elements around them with knee-jerk responses. In the safety industry, the only way to improve continuously is to manage using proactive, leading indicators. In the next chapter we will discuss what proactive and leading indicators are, how they can be used, and why they are better indications of safety performance.

As limiting and misleading as results-oriented (trailing edge) measurements can be, we must still utilize these indicators as part

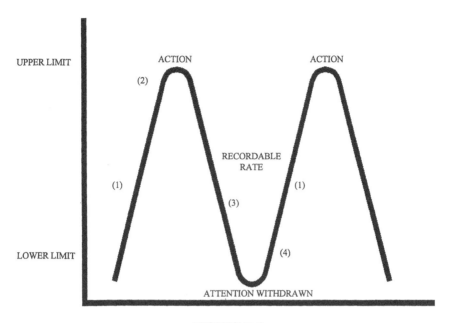

FIGURE 3.1

of our overall Business Measurement for Safety. These indicators should not be the entire basis for review of our safety performance. Results-oriented metrics are still valuable tools for internal and external safety performance measurement and comparison. Indicators such as OSHA recordables, lost time accidents, and severity rates all give valuable input into our safety performance. Additionally, they provide excellent benchmarking between companies as well as between industries, just as the number of widgets produced or pounds of production offer excellent benchmarks for production. An organization must measure these items only as a part of the safety-performance picture, being careful not to depend too heavily on results-oriented measurements.

NOTES AND REFERENCES

1. National Safety Council, *Accident Prevention Manual for Business and Industry, Administration and Programs*, 11th edition, 1997, p. 60.

2. Petersen, Dan, What Measures Should We Use, and Why? Measuring Safety System Effectiveness. *Professional Safety*, October 1998, p. 37–40.

3. Franklin, E. M., Injury rates don't tell the whole story, *Safety and Health*, National Safety Council, August 1998, p. 62.

4. Krause, Thomas R., *Employee-Driven Systems for Safe Behavior—Integrating Behavioral and Statistical Methodologies*, Van Nostrand Reinhold, New York, p. 4.

4

Behavior-Based
and
Proactive Measurements

Without measurement, accountability becomes an empty and meaningless concept.

Dan Petersen

When a company experiences a period of time in which no accidents or injuries occur, it is indeed tempting to think that safety performance has improved. Conversely, when accidents or injuries occur in proximity to one another, it is tempting to think that safety performance has somehow gone astray. Neither of these scenarios is necessarily true. In statistical terms, or in terms of an OSHA recordable rate, for a company with an annual incidence rate of 2.0, a quarterly rate of anywhere from 0.0 to 4.0 has no statistical value or meaning. Still, these are the very numbers that drive most safety programs today. Company management more often than not is predominantly motivated by these incidence rates on a monthly or quarterly basis. Many companies fall victim to the statistical smoke and mirrors of calculating OSHA recordables on a calendar basis. In other words, a company with an incidence rate of 14.7 on December 15 can easily show an incidence rate of 0.0 on January 15—hardly an accurate assessment of its true safety performance.

Figure 4.1 graphically shows the incidents that trailing indicators

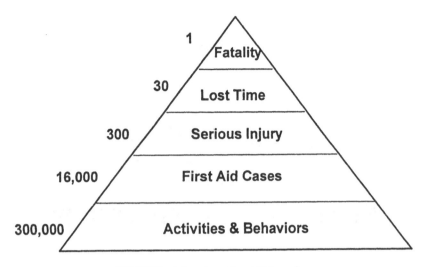

FIGURE 4.1 Accident Triangle.

measure, and the incidents that leading indicators measure. Behavior-based and proactive measurement must continually strive to measure activities and behaviors at the bottom of the accident triangle and do less reacting to the incidents at the top of the accident triangle.

With this type of numerical inefficiency, true progress in safety performance must be measured by some other means. Behavior-based safety is one performance measurement that must be considered. This method of safety-performance evaluation does not focus as intently on the results, but rather at the behaviors—day by day, week by week. In fact, Behavior-based safety performance measurement is fully capable of giving accurate, meaningful feedback on a minute-by-minute basis. Safety performance measurement moves from reacting to what *has happened* to managing what *is happening*. Behavior-based safety-performance measurement is focused on the people side of safety, not the physical hazards traditionally associated with accident prevention. Naturally, neither of these concepts of safety management is 100% pure physical hazard reduction or 100% pure behavior-based safety. Both must cross over to the other from time to time; essentially, they are two drastically different methods of management. Dan Petersen would even offer these words of caution concerning a program based totally on behavior-based criteria:

"I've seen the behavior-based approaches work beautifully

(even magically, in Scott Geller's words) in some organizations. I've seen these same approaches become a total disaster in other companies. In short, Behavior-Based Safety is no magic pill, no silver bullet. It can be helpful, even useful, in some organizations where it fits, and can be extremely counterproductive in those organizations where it does not fit."[1] Scott Geller reinforces this in an article in the January 1999 issue of *Industrial Safety & Hygiene News*. Geller says, "In fact, there is a 'magic' involved. Behavior-based safety stimulates and facilitates interdependent teamwork, which leads to innovation and creative synergy. To watch this transformation at work is a magical process. But this magic does not come easily nor quickly. It happens with proper top-down support and bottom-up involvement. Expecting too much too soon from behavior-based safety can result in disappointment and a label of 'failure'."[2]

Some believe the traditional injury frequency rates have no value because they rely on reactive, historical measures rather than proactive approaches to safety. Another criticism is that these statistics encourage management to react without providing accurate feedback on the effectiveness of the safety effort. Other experts believe that employee behavior is the most valuable indicator of safety performance. They say that if organizations can identify safety-related behaviors, they will have the only tool needed to improve workplace safety. Despite their differences of opinion on a standard of measurement (metric), most experts agree that management commitment is the principal determinant of safety performance. The bottom line is that all these measures are important.[3]

Research has proven over and over again that improving safety performance is never as easy as a one-step solution. There is no magic pill, no secret formula, no "ten steps to an accident-free world." Safety is a part of everyday life, which by definition makes it a very complicated process. Any safety performance improvement scheme that claims to have the necessary criteria to eliminate accidents once and for all is simply misguided. Throughout the remaining chapters of this book, many lists, elements, and criteria will be provided. They are supplied by some of the most highly respected safety experts in the field, as well as research provided by first-class universities and institutes. Regrettably, none will offer the solution; however, as a whole, they offer a wealth of information that will enable you to develop a safety-performance measurement tool that is ideally suited for your organization. Review each of the lists or elements given in any particular program with the thought of using the part of that program that best suits your company's needs.

The list of proactive, behavior-based (leading edge) indicators is as lengthy as that of any other business sector's performance standards. Management involvement and employee empowerment are just two proactive and leading edge indicators that are easy to track. Items that have been around for years but never used adequately for tracking performance include housekeeping, safety meetings, and audits. Keep in mind, once we begin, that to measure a specific metric, most often that metric will improve. Left unmeasured, the metric will often not be considered important. As Petersen states, "To hold someone accountable, we must know whether he or she is performing well, so we must measure that person's performance. Without measurement, accountability becomes an empty and meaningless concept."[4]

The list of metrics is only limited by the imagination and the needs of the company endeavoring to improve. More aggressive and advanced companies may include metrics using statistical process control, employee perceptions surveys, and the use of root cause analysis. Previously considered too subjective to be considered as part of the safety performance review, metrics can be quantified and used to advance the safety culture of a company. These would include: employee behaviors, safety culture, one-on-one safety observations, and many others.[5]

A behavior-based approach addresses, head-on, issues such as employee involvement, employee empowerment, the organization's culture, continuous training, and constant feedback. While these are just a sampling of behavior-based concerns, they are significantly different from a list of issues that are addressed using a physical hazard abatement approach. A traditional safety program concerned with abating hazards looks for things that have become a problem and then seeks ways to remedy the hazard. This type of approach to safety has been tremendously successful over the last few decades, but it appears that safety-performance improvement has plateaued. The low-hanging fruit has, for the most part, already been collected.

In a behavior-based approach, the underlying reasons for problems are sought more intently than the problems themselves. An example would be concentrating on fixing a flat tire instead of putting air in the tire each morning. Once the tire is fixed, the reason for the tire being flat should be considered. Once the reason the tire became flat has been corrected, preventing it from happening again is addressed. Once prevention of future flats is in place, the information that has been learned is disseminated to others in the organization to prevent them from having flats. This process doesn't stop until the whole process is reviewed to make sure that what was fixed, changed, or implemented is continuing to work as planned, and everyone who could benefit from the learned information is equipped with the needed information and equipment to benefit from the shared learning.

It might be worthwhile to consider some things that behavior-based safety is not. Scott Geller, in an article published in the January 1999 issue of *Industrial Safety & Hygiene News,* suggests the following:

Ten Common Myths about Behavior-Based Safety

- It's only common sense.
- It's just a fad.
- It's a magic bullet.
- Employees get the blame.
- It's only observation and feedback.
- Management is off the hook
- Environmental fixes aren't needed.
- It's 'touchy-feely' psychology.
- Attitude change must come first.
- There's no bottom-line payoff.[6]

Behavior-based performance cannot simply consider the question, "Did anyone get hurt?" That may be how management has viewed safety performance for some decades now, but increased competition will not allow this type of thinking to continue any longer. In fact, there are many forces that make this mindset a dangerous way of management in today's business world. Just a brief listing of some of the pressures contributing to increasing the necessity of running safety as a business sector might include:

- The liability for neglectful management is phenomenal.
- Downsizing has made every employee a critical link in the production process.
- No longer can injured employees be easily replaced with employees of similar skill and training.
- The overall competitive nature of industry places pressure in every area to provide production at maximum levels of achievement.
- Regulatory penalties for allowing unsafe conditions or practices to continue include personal liability for management and incredible sums of money in fines.

MANAGEMENT'S ROLE

Performance management must be driven by top management and linked to the business planning process. Performance management becomes the way in which strategic change is achieved, new cultures are built, and business initiatives such as quality improvement (Safety Improvement) and customer service are turned from ideas into reality.[7] Safety management must follow business management. The foreman or supervisor is still regarded as the key person for the implementation of safety. This continues despite business moving away from having any supervisors at all. Management is reaping the benefits of employee empowerment in other business sectors, while in the safety sector the classical management style of "management decides and the people follow" still dominates.

Strategic change cannot come from reacting to accidents and injuries. Strategic planning is proactive by its very nature. It is by strategic thinking that leading-edge indicators and proactive measurement can be derived. Thomas Krause, in his book *Employee-Driven Systems for Safe Behavior,* states it like this:

In the last analysis, the success or failure of Quality (Safety) initiatives does not depend on the brilliance or truth of the insights of Deming, Juran, and others. Whether in safety or in quality, success of these methods rest with leadership.

The critical test in industry now is how well managers are able to engage and sustain employee involvement in this new kind of applied science. Such ongoing engagement by the workforce is the driving mechanism of continuous improvement.[8]

In Chapter 3, the "Three E's" of safety—Engineering, Education, and Enforcement—were mentioned. While these have remained a dominant force in basic safety-program development, application, and enhancement, they no longer hold the entire picture of the modern-day safety system. Scott Geller, in his book *The Psychology of Safety—How to Improve Behaviors and Attitudes on the Job*, offers up three new "E's" for the modern day safety program:

> Ergonomics. Empowerment. Evaluation. I certainly don't suggest abandoning tradition — we need to maintain a focus on engineering, education, and enforcement strategies. But, to go beyond current plateaus and reach new heights in safety excellence, we must attend more competently to the psychology of injury prevention.[9]

We must be more aggressive in our management of day-to-day behaviors and activities. In fact, we must become experts at encouraging and promoting safe behavior. At the same time there are nonmeasurable functions that we must consider as well. Quoting from Geller again:

> Dr. Deming (1991, 1992) explained there are many things we should do for continuous improvement without attempting to measure their impact. We shouldn't do these things only to influence performance indicators, but because they are the right things to do for people. You might never be able to measure the impact of treating an employee with respect and dignity, but you do it anyway. Such treatment may, in fact, contribute to achieving a

Total Safety Culture but you'll never know it. Likewise, you'll never know how many injuries you prevent with proactively caring behaviors, and you'll never know how much actively caring behavior you'll promote by taking even small steps to increase coworkers' self-esteem, empowerment, and belongingness. You need to continue doing these things anyway. Many things that can't be measured and rewarded still need to get done.[10]

So how do we deal with all these different twists? On one hand we have results-oriented, reactive, physical hazard-types of measures on the safety system. They have worked well for decades and no one can dispute that they have improved our overall safety performance from, let's say, 1920 to the 1980s. But now this approach has ceased to give acceptable results, competition squeezes from every imaginable angle, the costs of injuries have skyrocketed to unbelievable heights, and all other business sectors are equally pressed for efficiency. Behavior-based approaches are contrary to traditional management strategies and beliefs. Proactive initiatives take commitment far beyond management's previous participation. The issues seeming to need tracking, measuring, and attention are "soft" issues. Previously, an orderly list of hazard abatement items could fit neatly into the overall business plan and strategies of the company. Now issues like industrial hygiene, ergonomics, carpal tunnel, and paradigms are circulating at every turn. How does a company go from tracking who gets hurt and little else to being actively involved in managing issues like the sense of belongingness, cultures, and empowerment?

WHERE DO WE START?

These issues, such as company culture, for example, are often viewed as "soft" issues that are difficult, if not impossible, to articulate. It is equally difficult for management to agree on just how to go about changing these soft issues. Business Measurements for Safety Performance provides a method in which these issues can be viewed, tracked, and managed just as any other "hard" component of business. It becomes rather simple to measure a wide range of safety issues in the same terms as other business sectors.

Scott Geller, in his article in the January 1999 issue of *Occupational Health & Safety*, "Behavior-Based Safety: Confusion, Controversy, and Clarification," outlined seven basic principles that should serve as a guideline when developing a behavior-based process for safety management.

1. Focus intervention on behavior.
2. Look for external factors to understand and improve behavior.
3. Direct behavior with activators or events antecedent to behavior, and motivate behavior with consequences or events that follow behavior.
4. Focus on positive consequences to motivate behavior.
5. Apply the scientific method to improve behavioral interventions.
6. Use theory to integrate information, not to limit possibilities.
7. Design interventions with consideration of internal feelings and attitudes.

These behavior-based safety principles are not a matter of opinion or common sense. They are based on behavioral science research and theory, developed to improve the performance of individuals, groups, organizations, and entire communities.[11]

For most organizations, accepting these new measurements is a big step. It often proves to be an even bigger step to actually implement them into a day-to-day management culture. Managers will need to manage in ways that improve the effectiveness of their staffs by the development of motivating work climates. To facilitate this change, companies need to break free from the bureaucratic and emotional restrictions of the "performance appraisal" and create a strategic lever that's every bit as powerful as the balance sheet and the organizational chart.[12] Before safety can become part of the overall strategy of an organization, it must be measured in the same ways that other business sectors are measured. Safety will not move from the plateau of past performance until the measurements of the past are coupled with the measurement methods of today and the future.

NOTES AND REFERENCES

1. Petersen, Dan, Behavior-Based Safety: Build a Culture or Attack Behavior? *Occupational Hazards*, January 1999, p. 29–32.

2. Geller, Scott, The Ten Myths of Behavior-Based Safety, *Industrial Safety & Hygiene News*, January 1999, pp. 14-16.
3. Herbert, David A., How to measure where your organization has been, where it's at and where it's going, *OH&S Canada*, March/April 1995, p. 54-60.
4. Petersen, Dan C., *The Challenge of Change, Safety Training Systems*, Creative Media Development, Inc. 1993.
5. O'Brien, Dan, Where Do You Fit In? Safety Program Stages, *Industrial Safety & Hygiene News*, Vol. 29, No. 10, Oct. 1995.
6. Geller, Scott, The Ten Myths of Behavior-Based Safety, *Industrial Safety & Hygiene News*, January 1999, pp. 14-16.
7. Weiss, Tracey and Hartle, Franklin, *Reengineering Performance Management—Breakthroughs in Achieving Strategy through People*, St. Lucie Press, Boca Raton, FL, 1997, p. 8.
8. Krause, Thomas, *Employee-Driven Systems for Safe Behavior—Integrating Behavioral and Statistical Methodologies*, Van Nostrand Reinhold, New York, p. 4.
9. Geller, E. Scott, *The Psychology of Safety—How to Improve Behaviors and Attitudes on the Job*, Chilton Book Company, Radnor, PA, 1996, p. 27.
10. Geller, E. Scott, *The Psychology of Safety—How to Improve Behaviors and Attitudes on the Job*, Chilton Book Company, Radnor, PA, 1996, p. 334.
11. Geller, Scott, Behavior-Based Safety: Confusion, Controversy, and Clarification, *Occupational Health & Safety*, January 1999, p. 40-49.
12. Weiss, Tracey and Hartle, Franklin, *Reengineering Performance Management—Breakthroughs in Achieving Strategy through People*, St. Lucie Press, Boca Raton, FL, 1997, p. 22.

5

What Other Business Sectors Measure

No problem can be solved from the same consciousness that created it.

Albert Einstein

When Americans saw the quality of our goods surpassed by that of foreign countries, American industry as a whole had to rethink its commitment to quality. It wasn't that quality wasn't important or that America turned out bad products. The reality was that to stay competitive, America had to get into the game. Quality had to become part of the culture. Today, safety must be inserted into our industrial culture in the same way and with the same ferocity that quality has been. Just as quality had to be raised to a higher level—it had to become a value—safety must become a value for our employees, our managers, our owners, and even our families.

Increasing productivity and quality is not a neatly packaged problem. It is a messy, poorly defined problem with no beginning and no end—it is difficult to decide where to start.[1] The same thing can be said of the safety movement. Once an organization moves away from the false comfort of incidence rates and other trailing indicators, the problem does, to some degree, seem like a poorly defined problem with no beginning and no end. An organization need not look far to find well-defined, thoroughly tested, and reality proven methods of safety system measurement. An organization need only look to the other business sectors within its own organization.

The following takes a quick look at some simple examples of what other business sectors might consider in measuring their performance. Remember, these are just a handful of examples in a virtually endless panorama of information and behaviors that can be, and most often are, measured to evaluate performance.

What Business Sectors Measure

Sales

Percent of orders with less than normal lead time
Order process time
Cost per dollar of sales volume
Percent increase over previous time period

Engineering

Time for drawings to be released
Number of changes required on specifications
Production delays attributed to Engineering
Percent of budget overruns

Materials Management

Percentage of on-time purchase deliveries
Percent of inventory over ideal
Cost savings
Optimal inventory management

Maintenance

Minimize amount of downtime due to maintenance problems
Percent of predicted equipment maintenance
Percent of preventative maintenance preformed before failure
Work order backlog

Customer Service

Percent of phone calls picked up within thirty seconds
Percent of complaints answered within one day
Decrease percent of inquiries requiring two or more contacts
Cost per order

Human Resources

Dollars spent for recruitment
Number of labor disputes
Average expense per new employee
Average employee turnover rate

Production

Quality of product
Percent of on-time production
Percent of offspec material
Manhours per product produced

In his book *World Class Manufacturing—The Lessons of Simplicity Applied*, Richard Schonberger explains world-class manufacturing as:

> A full range of elements of production are afforded: management of quality, job classifications, labor, training, staff support, sourcing, supplier and customer relations, product design, plant organization, scheduling, inventory management, transportation, handling, equipment selection, equipment maintenance, the product line, the accounting system, the role of the computer, automation, and others.[2]

Every aspect of our business is measured and controlled. The extent to how well we succeed in business is largely dependent on how well we manage these many aspects of our business. The quality movement in America has provided exponential returns because industry learned how to control what industry previously either didn't feel was important enough to control or didn't know how to control.

The number-one objective of any business is profit. In business profit is number one. That doesn't mean that production is unimportant. Nor does it mean that safety cannot receive the proper amount of attention and effort. For an organization not to be focused on profit is for an organization to not understand what keeps the doors open—what pays employees' salaries; what provides capital for equipment, offices, and other necessities. I cringe whenever I see a safety banner that says "Safety First." On the surface that sounds great, but in reality it will never work. Safety doesn't need to be first. It needs to be an integral part of everything. Safety must be an integral part of the first, the last, and everything in between. As the quality movement continues to not only grow but become a value of modern industry, it becomes clearer that a critical key is "Do it right the first time." How well this fits with the safety movement. The first time is the only time we have in safety. I'm sure many an accident victim wishes he or she had another opportunity to do it right!

Knowing that the answer to, "What do other business sectors measure?" is, in a word, "Everything." It becomes a little easier to move away from trailing indicators and focus on where an organization needs to focus. The question becomes not what other business sectors are measuring, rather, that we use the same philosophy to measure safety as is used to measure production, sales, and all the other business sectors. There should be no difference between how we measure other business sectors and how we measure the safety system. Once those walls are broken down and removed, the entire business world is opened up to safety-performance measurement. It's that simple. In order for safety performance to improve, we have to measure the areas of safety performance that we want to improve. Regardless what the metric is, if we want it to improve it, we must measure it. Left unmeasured, the metric will most likely remain unimproved.

This approach also makes it easy to incorporate "management" versions of a safety approach at the same time as "behavioral" approaches. There ceases to be a need to have the management approach square-off with the behavioral approach. The two, and even others, can fit very nicely into a package that is right for your organization. Statistics and track records support many different approaches to the safety performance measurement issue. Again, there is no magic bullet, there is no magic pill, and there is no one thing that will put accidents behind your organization forever.

With so much needing to be measured, how does an organization decide what to measure? There are several basic areas to look at to determine that:

- Areas within your organization that obviously need improvement
- Areas of past or possible significant loss
- Areas where constant improvement is desired
- Areas observed in other organizations in which similar results are desired

A beneficial method of determining exactly *what* should be measured, and *how* it is to be measured, is to benchmark an area that is already performing at desired levels. Because this is such a beneficial tool, we will take a more in-depth look at benchmarking.

Benchmarking can easily be thrown into the hat with other industry buzzwords like Behavior Modification, Paradigm Shifts, and Globalization. In a decade where Yuppies are intensely concerned with whether their cars measure up to the neighbors', it should be no surprise to us that industries have become very interested in comparing themselves with other industries. Thus, enter in the newest buzzword in industry: *Benchmarking*. Simply put, benchmarking is the comparison between one company's performance in a particular area and another company's performance in that same area. Benchmarking sounds simple enough, but all too often we are quick to offer comparisons with the Monsantos, Procter & Gambles, Occidental Chemical Corporations, and Dow Chemicals of the world. This is a commendable aspiration, but most likely not very accurate or helpful. In fact, it most probably is like comparing your son's little league pitcher to Nolan Ryan— very flattering, but useless.

Attempts to benchmark with another company or industry should be carefully thought out and organized before any step toward comparison is taken. For example, it would serve no useful purpose for ACME Maintenance Company to benchmark Procter & Gamble. While the differences are obvious (i.e., company size, products, and market share), there are no real connecting points between the two companies. This doesn't mean that Procter & Gamble doesn't have anything from which ACME can learn. It's just not the most productive tool for improvement on Acme's part.

Selection in the benchmarking process should include several criteria in order to obtain successful comparisons. The first consideration should be to determine exactly what is desired from the benchmarking. In most cases comparisons in a very narrow field of evaluation will be most helpful, and probably all that can be utilized for benefit at one time anyway. For example, collecting benchmarking data for incentive programs may be beneficial, whereas data on OSHA recordables may not be comparative or useful. The point is, to benchmark "the safety program" will most likely only confuse and frustrate the recipient of the data. To benchmark a safety program would be similar to asking an auto parts person if he or she had any Fords in stock. Benchmarking should be used to initiate or improve a specific sector of your safety program. Some examples of benchmark sectors that might be of a helpful nature would be: job hazard analysis utilization, incentive programs, employee observation programs, accident investigation techniques, return to work policies, employee rotation, and ergonomic program implementation, among others.

Benchmarking is a critical link in today's business environment because of the incredibly competitive nature of the marketplace as a whole. Companies are forced to produce more end product with fewer people, and at a faster and more consistent pace. Competition is on a global basis, which only increases the need for "best of the best" practices. Constant improvement in every area is essential to remain competitive on a global basis. Benchmarking allows program development in a specific area to take place in less time, with less expenditure, and all on a shorter learning curve. Benchmarking allows companies to heighten their market awareness with real-time, actual situations. It lessens the need to reinvent the wheel. Benchmarking allows learning across industry, company, and geographical boundary lines.

We must benchmark the specific leaders in industry that excel in the specific area in which we want to improve or implement new programs. They might be called "Niche Heroes." Programs like Rohm and Haas' home safety program, Lockheed's procedures awareness and employee interaction program, General Electric and American Ref-Fuel's VPP programs. These are companies that have moved to the top of a particular part of the safety culture. They possess specific areas of excellence. In many cases these will be industry leaders, but most often they will be companies that have had all the elements of success come together in just the right mix. In some cases the companies to benchmark have come to excellence through setbacks, or even catastrophes. An example would be Phillips Petroleum's Sweeny, Texas plant. After its 1989 explosion, the plant developed one of the best contractor interfaces in the industry.

Most benchmarking is done over the phone. If you are benchmarking major programs there may be a need for a site visit, but most of the time a simple chain of phone calls can get the information you need. Contacts through professional organizations make this step much easier. Involvement in the local chapter of the American Society of Safety Engineers (ASSE), American Industrial Hygiene Association (AIHA), and other local, state, and national safety and health organizations prove to be excellent opportunities to find and develop contacts worthy of benchmarking. Attendance at annual professional development conferences such as those sanctioned by ASSE, American Conference of Industrial Hygienists (ACIH), and Voluntary Protection Program Participants Association (VPPPA) all add to your window of opportunity to benchmark the best in industry.

A common place to start the benchmarking process is with your company's competitors. This can be a good place to begin gathering information because there will probably be some natural similarities in size, equipment, cultures, etc. Starting with your competitor may have some drawbacks. First, there may be more reluctance to share data because of fears of antitrust litigation. While this is a genuine concern, proper dialogue will avoid any conflicts in this area. Remember that you are seeking safety and health-related information, not process design, capacities, or any proprietary information. Your contacts will probably be with the safety and health manager, and issues of trade secrets and formulas should never come up. It they do come up, avoid them at all costs and end the contact immediately. In most cases safety managers are more than willing to share their success stories with you, whether they are competitors or not. If you have properly defined what area you want to benchmark, concerns of proprietary information shouldn't be an issue.

Be considerate of the person you are asking to help you benchmark. Typically, a phone call requesting information on benchmarking "your return to work policy" or benchmarking "your hoist inspection program" are easy to explain, not considered confidential, and don't take long to collect the necessary data. Keep your scope of benchmarking narrow and specific so as to not take up large amounts of time from the person supplying the information. It's also important to maintain the same benchmark with each contact you make. It's easy for the scope of the benchmarking to grow wider as you accumulate information. In some cases your request to acquire benchmarking data may be turned down. While I haven't found this to be very common, if it happens just go on to the next company.

It is customary for you to share what your company is currently doing in the benchmark area. This can be done up-front and often relaxes any concerns the other person might have because you're already sharing with them. Most companies that share benchmarking information will expect to receive some kind of report or follow-up information when you are finished with your benchmarking. Follow-through in this area is critical because failure to fulfill your end of the bargain could nullify your ability to obtain benchmarking information in the future.

The pitfalls in benchmarking can be much like following the wrong road on a road map. It's always best to be sure you're on the right highway before speeding off into the sunset. Often in some company's quest for greatness, benchmarks are often set too high, setting the company up for eventual failure. In picking benchmarking standards, a company must pick benchmarks that are similar to existing needs, are meaningful, and are high enough to require effort but are within reach for employees and management to strive for with reasonable expectation of achieving.

Benchmarks that are too broad in scope can also be detrimental to a company. Seldom would one company seek to benchmark every program a mentor company might have. More productive would be selection of specific programs that closely fit the company's culture and abilities. When specific goals are set, employees and management can picture attainment of the predetermined benchmark, and achievement of the benchmark is much more realistic and tangible.

Another pitfall of benchmarking is pushing the benchmark harder than the foundations of the safety program. In other words, we must keep the "main thing" the "main thing." When we lose sight of our safety programs in pursuit of a benchmark goal, we end up losing ground on the safety. Nothing is more wasteful than doing with great efficiency that which is totally unnecessary.[3] Given these general parameters, benchmarking can be the tool of choice for companies to choose the "best of the best" for constant improvements in specific areas of safety and health programs. By looking at better safety systems components we can gain insight as to some of the keys to safety excellence.

Now that we've seen how we can look at our organizations to see what we need to improve and couple that with looking at other organizations that are excelling in that particular area, let's discuss how we get started with some sort of meaningful system. Measurement of these areas does not have to be rocket science and does not have to be a cumbersome or lengthy exercise. An important concept to insert at this point is that it is absolutely critical that Business Measurements for Safety Performance does not become the main thing. The main thing is improving the safety performance. Business Measurements for Safety Performance is only a tool. It is imperative that it be kept simple, understandable, and easy to administer.

The first step is to decide where to start. Whether you want a comprehensive system or just want to get your feet wet with Business Measurements for Safety Performance, a measurement system can be devised for your organization. A preliminary stage in the safety system analysis is to find out where exactly your organization is. Basic information gathering must take place. What currently exists in the organization? Most often, creating a checklist can do this . Whether original or borrowed, fundamental elements are easy enough to list and then check off if the organization is performing those functions. Compose a list of metrics that are important to your company and that you want to see improved. Table 5.1 shows an elementary system, while Table 5.2 serves as an example of a more comprehensive system. Both are given only as examples. Again, you will have to evaluate what your organization's needs are, see what works for your organization, and decide what fits best with your organization's safety culture.

TABLE 5.1   Basic Metrics Checklist for Your Company

Metric	Points
OSHA recordable rate	10 points
Severity rate	10 points
Insurance reserves	10 Points
Safety meetings	10 Points
Safety observations	10 Points
Safety audits	10 Points
Safety incentives	10 Points
Housekeeping	10 Points
Documentation	10 Points
Management involvement	10 Points
Total — All Metrics	**100 Points**

TABLE 5.2   Comprehensive Metrics Checklist

Metric	Points
OSHA recordable	5 points
Severity rate	5 points
Insurance reserves	5 Points
Safety meetings	5 Points
Safety observations	5 Points
Safety audits	5 Points
Safety incentives	5 Points
Housekeeping	5 Points
Documentation	5 Points
Management involvement	5 Points
Contractor severity rates	5 Points
One-on-one safety observations	5 Points
Root cause analysis	5 Points
Compliance	5 Points
Employee empowerment	5 Points
Contractor interface	5 Points
Employee behaviors	5 Points
Employee perceptions	5 Points
Overall safety culture	5 Points
Statistical process control	5 Points
Total - All Metrics	**100 Points**

Notice that in both Table 5.1 (an elementary system) and Table 5.2 (a more comprehensive system), the metrics used to complete the system are equally weighted. For example, the system weight for Safety Incentives is equal to the system weight for, let's say, Housekeeping. The indication here is not that Safety Incentives and Housekeeping are equal in importance; rather, that both have been predetermined to be integral pieces of the safety system for your particular company. Both have been designated as areas where continuous improvement is needed or areas where excellence has been achieved and a method of monitoring the sustaining of that excellence is developed. This method of equal weighting is not essential for the Business Measurements for Safety Performance system to work, but it makes the system much easier to design as well as to implement and track. Once uneven weight is introduced to the system, all sorts of subjective pressures begin to distort it.

One might compare Business Measurements for Safety Performance to an automobile. Is the tire more important than the wheel? Or is the brake more important that the accelerator? In Biblical terms, "The eye cannot say to the hand, 'I have no need of you;' nor again the head to the feet, 'I have no need of you;' indeed, those members of the body which seem to be weaker are necessary."[4] Business Measurements for Safety looks at the whole safety system, understanding that there are many important parts and little value is achieved by subjectively weighting one sector heavier than another.

Keep in mind that neither of the systems shown (Table 5.1 or Table 5.2) are the ultimate Business Measurements for Safety Performance. Your organization's parameters may be a blend between the two, or may have metrics completely unique to your organization. Your metrics will not be benchmarks between companies or industries as are some trailing edge indicators such as OSHA recordables. The metrics you choose may not even provide a good benchmark between plants within the same company, although similar metrics may be helpful in the beginning. The focus is constant improvement of leading edge indicators that will ultimately improve the trailing indicators.

NOTES AND REFERENCES

1. Kaydos, Will J., *Measuring, Managing, and Maximizing Performance*, Productivity Press, Cambridge, MA, 1991, p. xvi.
2. Schonberger, Richard, *World Class Manufacturing—The Lessons of Simplicity Applied*, The Free Press, New York, 1986, p. 1.
3. Kaydos, Will J., *Measuring, Managing, and Maximizing Performance*, Productivity Press, Cambridge, MA, 1991, p. 17.
4. *New King James Version of the Holy Bible:* 1 Corinthians 12:20–21.

6

Quantitative vs. Subjective Measurement

Accountability without measurement is worthless.

Daniel Patrick O'Brien

Let's begin our examination of quantitative vs. subjective measurements with explaining a basic premise on which we must build our measurement beliefs and our measurement systems. Everything—everything!—is measurable. Moreover, everything that is measurable can be quantified. We'll discuss this point more thoroughly, but that's the basic premise upon which all measurement, safety or otherwise, must be based. Measurement does not necessarily mean percentages, number of feet, number of pounds, or other specific metrics. In fact, some of the "dyed in the wool," old standby quantitative numbers seem to crumble when scrutinized. For example: "a supervisor of ten people can do absolutely nothing all year and attain a 0.0 frequency rate. Using a statistical measure to judge his or her performance is ludicrous."[1] An organization can have a horrendous injury frequency rate on December 15th yet have outstanding frequency rates on January 15th. So with just this simple example, the old standby—the OSHA recordable incidence rate—doesn't seem quite so "matter-of-fact" as it would first appear.

Industry as a whole has fallen victim to widespread and total dependency on the OSHA recordable incidence rate as the sole judge of safety-performance measurement when, in reality, it is little more than a rough indicator of many systems and cultures that are more complex. At what point do accidents become a valid measure by which to judge performance? Actuaries state that only after about 1,129 accidents have occurred would they begin to judge the unit that generated those 1,129 to be so believable that future rates could be based on those figures.[2] These are startling statistics when most organizations today may not have that number of incidents in decades, if not centuries. Yet, industry trudges on using these indicators primarily because they are comparable with other industries and organizations. These crude indicators also offer a thoroughly read "how to" manual produced by the U.S. Department of Labor, O.M.B. No. 1220-0029, "Recordkeeping Guidelines for Occupational Injuries and Illnesses." This makes it simple (some may disagree just how simple) for those who want little more than to comply with government regulations and accept the current level of safety nonperformance.

Measurement of company morale would most often fall right at the top of a company's "subjective" measurement category. Issues such as morale are often considered so subjective that the entire metric is thrown out along with all of its potential benefit, never to be considered again as an indicator of performance. The bottom line result being frustrated employees getting more frustrated, and management continuing to look for some magic pill that will fix all of their problems in one quick, easy step, all the time wondering why "trailing indicators" continue to show negative results. There are many ways to take this historically "subjective" measurement and extract meaningful, beneficial, and, most of all, quantitative information. For example, the rate of turnover and amount of absenteeism may indicate low morale, just to mention two simple approaches. Both of these examples can easily be quantified.

With this general "everything is measurable" approach, a whole new world awaits the manager who wishes to measure ALL the things that effect a business sector. This is not such a new fangled concept for most business sectors. Production, quality, shipping, sales, and most others figured out a long time ago that performance measurement must include more than just the bottom line if there is ever going to be hope for constant improvement. The safety business sector has been slow to embrace the concept that everything is measurable and that if it doesn't get measured, it probably won't get done. Most other business sectors have not only embraced the "what gets measured, gets done" concept, they have learned to rely on it as a core reality of performance measurement.

A good performance measurement system, regardless of whether it is subjective or quantitative, consists of leading indicators or trailing indicators and is developed, not designed. Yes, it does need significant "design" to obtain proper criteria, focus, and direction. However, it is past that where the individualism of each organization must take over and mold the "designed" system into a finely tuned and "developed" system that matches the culture of that particular organization. In other words, all safety performance measurement systems must have some core focuses and values, but what works very well for one organization may not work at all for another.

Selecting good measures is neither obvious nor simple. Qualitative research is always an output of the fruitfulness of a human mind, both in generating (subjective) hypotheses and in being insightful enough to select measures that will test the hypotheses.[3] This is where the frequently stated line, "there is no magic pill for safety improvement," comes into play. When a company stops looking for the magic pill and starts looking inwardly in search of ways that constant improvement can take place, trailing indicators suddenly become less important. Metrics that truly need to be improved can be brought to the surface and dealt with.

Statistical analysis tools have become essential for many sciences. Statistical texts often imply that data analysis is purely mechanical; practice suggests that statistical analysis is a judgmental process with many subjective steps.[4] While Muckler's hard line here on statistical approaches may seem to degrade the importance of good statistical data, I've only included it in an attempt to illustrate some simple points.

1. It warns one to be particularly sensitive to the possibility of subjectivity, to be especially aware of judgmental steps in measurement, and to take particular care in the collection, analysis, and interpretation of data.

2. Statistical data can easily become a numbers game. Data that started out to be quantitative and very straightforward often ends up very manipulated and, thus, meaningless at best.

3. One of the most common pitfalls of organizations is to become "statistics maniacs." Statistics can be very valuable. They can be indispensable. Statistics can also be the sand in the gears of a finely tuned organization. Part of the development of an organization's safety performance measurements must include a hard look at what numbers are really needed, and which ones will be beneficial in achieving constant improvement in the safety system.

The most important aspect of performance-measurement selection is that the information collected be evaluated to ensure that the metrics are useful and provide needed, valuable information, whether it be quantitative or subjective. It's important to establish a system that is structured around parameters that will be beneficial to your company. Input from the groups or departments that are to be measured should be considered and included in establishing the safety metrics to be used. Input from the employees in this area is critical. Its also important that employees understand that metrics that were previously perceived as subjective will now be measured in quantitative terms. Employees can play a valuable role in developing quantitative metrics for previously subjective parameters.

Selection of safety-performance metrics need not be difficult, although it should go through some logical thought processes and meet some basic criteria. Ultimately, the metrics developed will be used to hold individuals, departments, managers, and entire organizations accountable for their safety performance. To hold someone accountable, we must know whether he or she is performing the very things for which he or she is being held accountable. Accountability without measurement is worthless. For most everyone—line workers, supervisors, managers, even executives—to measure is to motivate.

Let us look at some basic, very preliminary criteria an organization might look at right out of the starting blocks to ensure that safety performance measurement criteria get off on the right foot. Meister (1986) proposed 14 specific criteria for selecting performance measures:[5]

Meister's 14 Points

1. Directly relevant to the output
2. Directly observable in task performance
3. Not requiring additional interpretation
4. Reflecting critical task events
5. Precisely definable
6. Objective
7. Quantitative
8. Unobtrusive
9. Easily collected
10. Without specialized instrumentation requirements
11. At appropriate levels
12. As inexpensive as possible
13. Reliable
14. Meaningful to researchers and decision makers

Meister's 14 points by no means rule out the use of subjective data to evaluate safety performance. In fact, a closer look reinforces the belief that all things, subjective and quantitative, should be equally evaluated for benefit in the safety performance measurement system. Furthermore, so much data is supportive that subjective data may indeed be some of the better data to start with. In answer to "Why collect subjective data?" Gamberale et al. responded this way:

> An evaluation of the work environment from the worker's own point of view is in itself of obvious interest in almost all types of work environment studies since the subjective well-being of most activities is related to the work environment.[6]

A traditional objection to subjective methods is their alleged lack of reliability. However, there is overwhelming evidence that it is possible to measure subjective responses with good reliability. "An unreliable subjective measurement is more often the result of deficiencies in the investigator's competence, than of an inherent quality of these measures."[7]

Hopefully, by now you have realized the importance of tracking much more than just OSHA recordable or even lost time accidents. Some of these metrics, such as contractor incidence rates or insurance reserves, would be easy enough to quantitatively track. However, most would readily agree that tracking other performance metrics that might not be as easy to do quantitatively would be beneficial if a fair and consistent method existed. Performance metrics such as employee behaviors, housekeeping, one-on-one supervisor evaluations, perception surveys, employee involvement, employee contentment, and hundreds more could be used for beneficial safety-performance measurement. Obviously, some of these would be considered more as "core" safety performance issues, while others would be evaluated as peripheral safety performance issues.

Let us look at a simple example of how typically subjective metrics might be quantitatively measured.

Area Housekeeping
(Ten-point possible score – One point each)

☐ Is there a formal housekeeping audit program in place and being utilized?

☐ Are safety incentives or awards tied to housekeeping audit results?

☐ Do housekeeping audits address every area of the facility including office areas, break rooms, and locker areas?

☐ Are results from housekeeping audits documented and tracked to completion?

☐ Are housekeeping results conveyed to all employees?

☐ Are individual employees responsible for specific housekeeping duties?

☐ Is housekeeping considered before a job is completed?

☐ Are routine inspections conducted to ensure housekeeping on every job is completed before employees leave the job?

☐ Are facilitywide cleanup times used?

☐ Are trashcans or bins readily accessible throughout the facility?

The above shows how a basic safety metrics system might be evaluated. The scoring in the figure would be part of a ten-step Business Metrics for Safety Performance Measurement system. It shows how something as subjective as housekeeping scores can be put in predetermined and agreed to criteria that can be quantitatively measured for improving safety performance. This particular example would have nine other similar components, each worth another ten points, totaling 100 points. The higher the score, the more positive the overall indicator.

Several important points are worth reiterating. First, this is just a guide and must be customized for your specific safety culture. Tracking usage of Statistical Process Control methods and use of Root Cause Analysis will be of little value for a company in the early stages of development of its safety culture, whereas a company on the more advanced end of the safety culture may find less value in tracking whether or not basic safety meetings were conducted.[3] Secondly, it is important to include a mix of both leading and trailing edge indicators in your safety metrics. The metrics should be weighted more heavily with leading indicators. This is the area where you can make a difference with a proactive approach. It is in these leading edge areas that you can begin to home in on areas where constant improvement is important to your company.

As more advanced indicators are included, at least on the surface these indicators will appear to be more subjective. Because an indicator may have subjective parameters within it, the indicator does not necessarily have to be measured on a subjective basis. By predetermining the assigned value of a particular aspect of the individual metric, that parameter can become a quantitative measurement although the overall concept may be subjective.

Let us look specifically at a metric that might appear to be very subjective on the surface—an employee perception survey. Within a survey, questions may be asked that are subjective, biased, or misleading when viewed alone. Nevertheless, the Employee Perception Survey is a valuable tool that can easily be used in a quantitative way. Petersen states: "The perception survey is used to assess the current status of an organization's safety culture. Critical safety issues are rapidly identified and any differences between management and employees on the effectiveness of the company safety programs are clearly demonstrated."[2] The perception survey is too valuable a tool to be passed off quickly as too subjective. Surveys are commercially available that have been scientifically validated and provide meaningful and quantitative data. Their inclusion in your safety metrics can be the basis for initial growth in this area. The following shows how surveys can be quantitatively measured. This example is indicative of a program that is not currently utilizing an employee survey and wants to initiate continual improvement in that area.

Employee Perception Survey
(Ten-point possible score—Two points each)

☐ Has a statistically validated survey been conducted within the last 18 months?

☐ Have results from the survey been communicated to the employees?

☐ Has a follow-up program been implemented to address the areas of concern expressed on the perception survey?

☐ Have employee interviews been conducted to establish areas of concern?

☐ Has a team of employees been established and trained to conduct surveys and/or interviews to obtain valid information?

These examples show how very easily performance criteria and program areas can be placed in a system, allowing them to be quantitatively placed into a larger picture of overall safety-performance measurement. The only criteria are some basic and fundamental reviews of your organization to determine where the greatest need for improvement lies. Once areas of concern have been identified, specific objectives can be combined and agreed upon by all involved. Agreement must be across-the-board, including employees, supervisors, managers, and executives. Once everyone agrees on what will be measured and specific predetermined criteria for how it will be measured is set forth and communicated to all, it becomes easy to achieve improvement in areas of concern that were not previously addressed.

We'll see more examples in coming chapters, but for now let's move our focus to constant improvement of those areas that we have determined need attention in our organization.

NOTES AND REFERENCES

1. Petersen, Dan, *Analyzing Safety System Effectiveness*, 3rd edition, Van Nostrand Reinhold, New York, 1996, p. 15.
2. Petersen, Dan, *Analyzing Safety System Effectiveness*, 3rd edition, Van Nostrand Reinhold, New York, 1996, p. 15.
3. Muckler, Fredrick A., Selecting Performance Measures: "Objective" versus "Subjective" Measurement, *Human Factors*, August 1992, Vol. 34(4), p. 441–445.
4. Muckler, Fredrick A., Selecting Performance Measures: "Objective" versus "Subjective" Measurement, *Human Factors*, August 1992, Vol. 34(4), p. 441–445.
5. Meister, D. *Human Factors Testing and Evaluation*, Elsevier, Amsterdam, 1986.
6. Gamberale, Francessco, Kjellberg, Anders, and Akerstedt, Torbjorn, Behavioral and psychophysiological effects of the physical work environment—research strategies and measurement methods, *Scandinavian Journal of Work, Environment, and Health*, 1990, p. 5-16.
7. Gamberale, Francessco, Kjellberg, Anders, and Akerstedt, Torbjorn, Behavioral and psychophysiological effects of the physical work environment—research strategies and measurement methods, *Scandinavian Journal of Work, Environment, and Health*, 1990, p. 5-16.

7

Constant Improvement and the Safety Continuum

Sometimes leaders think they are doing their job just because there is a lot of hammering going on.

Laurie Beth Jones

Kaizen (Japanese for constant improvement) has become well-known to managers of companies over the last 15 years. The Constant Improvement Model was originally developed in Japan. Once proven there, the concept swept through the United States like wildfire in the 1980s. Total quality and continuous improvement are used to control products, equipment, zero-tolerance defect production, just-in-time production, employee involvement, consensus decision-making, and other management focuses widely used in most businesses today. Managers have learned that in *Kaizen*, as in any other management theory, to make a theory work in an organization they must adapt the theory to their organization, not adapt their organization to the theory. This same principle applies to safety as it applies to quality, production, and all other business sectors.

Our industry's leaders have watched the tremendous impact that constant improvement philosophies have made on our business, our culture, our economy, and everything else in our society. It is somewhat amazing that most have not yet applied these same time-proven philosophies to the safety industry. *Kaizen* focuses on continuous improvement where the focus is on processes rather than results. It is not that the results are unimportant; they are paramount. However, the results will take care of themselves if the processes are functioning in excellence. Hopefully, by now you have seen enough measurable safety processes to at least wonder whether the results (injury rates) wouldn't take care of themselves exactly as they do in quality, production, and the rest if all of those safety processes are performing at excellence.

The continuous improvement philosophy has merit in safety in exactly the same way it does in other business sectors. Key points may be expressed differently; the points may focus on slightly different areas and may tend to be more "subjective." The bottom line is that they affect safety performance identically in those sectors they were originally designed to improve (quality and production). The following shows the primary points of the *Kaizen* (continuous improvement) principles.

Principles that Can Be Applied to All Areas of Business, Including Safety.

- Focus on the customers
- Continuously make improvements
- Acknowledge problems openly
- Promote openness and honesty
- Create work teams
- Manage projects through cross-functional teams
- Nurture the right relationships process
- Develop self-discipline
- Inform every employee
- Empower every employee

All of these principles apply to the safety sector just as they would in quality, production, maintenance, or any other business sector. This approach sounds so simplistic. Why does industry have such an aversion to placing safety-performance measurement in the same management arena as all other business sectors? This question may be answered most clearly by explaining the basic safety continuum; where an organization lies on the continuum of safety development has a significant impact on how openly ideas like continuous improvement and quantitative measurement of subjective criteria are accepted.

All organizations, large and small, undeveloped and highly sophisticated, fall somewhere on the safety continuum. The following few pages may shine light on areas of your organization and allow you to see fundamental mindsets that have prevented progress on one hand, and guaranteed success on the other. Keep in mind that organizations do not usually fall neatly into an exact spot on the continuum; rather, some areas lag while others flourish.

THE SAFETY CONTINUUM

There are safety systems in every stage of development and implementation—some in the elementary stages of the safety culture and some on the cutting edge of safety thinking. Where an organization is on this continuum between elementary thinking and advanced thinking determines the quality and effectiveness of the overall safety system. An organization may or may not fit neatly into any one stage of development on this continuum. More likely than not, an organization would have characteristics both trailing and leading the stage of the continuum currently occupied.

The most rudimentary position on the safety system continuum is that of the organization that is totally unaware a safety culture even exists. In this state, evaluation of the safety aspects of an operation is nonexistent. Tasks are performed with no thought given to how safety fits into the operation. In this stage, employee injuries are considered as part of the operation. No consideration is given to proactive approaches in preventing accidents. What little consideration has been given to safety quickly returns the verdict of too costly, too much trouble, or simply not worth it.

To progress from the stage of being unaware of any safety culture requires development of a basic safety program. This entails written formalization of policies, development of standard operating procedures for equipment and processes, training on basic safety standards such as hazardous communication, lockout, personal protective equipment, confined space entry, etc. Injury/ illness data would be assembled, tracked continuously, and compared with comparable industries or businesses. Minimum standards for compliance with governmental regulations would be implemented. Employees would be required to report accidents and significant incidents. Improper techniques and systems would be identified and corrected. Safety would be involved in the engineering of projects, the purchasing process, human resource functions, and other sectors. Monetary funds and time would be allotted for safety initiatives and concerns. Participation in the safety program would become mandatory.

Advancing on the safety continuum, but still in the elementary stages, is the organization that functions out of responsibility. That is, employees respond to safety needs and problems out of a responsibility that has been given or dictated to them. This group would just as soon not have the responsibility of safety. It is burdensome to them, probably one of the most hated responsibilities they must deal with in their work environment. They see little or no value in safety activities. Their participation is forced from a higher level of authority. Upper management is aware there is a need for safety, but they really don't know how to deal with safety issues other than behind a facade of importance. Upper management may also feel somewhat burdened by safety issues and responsibilities. At this level, the safety manager would be responsible for all training, inspections, documentation, and permit writing, as well as all other safety activities. If the task involves safety, the safety manager would be solely responsible for the planning, implementation, and success or failure of the safety system.

Progression from this stage would include specific goals and objectives being established. All levels of the organization would be familiar with safety goals and objectives, and they would be consistently sought after and included in activities. Employee performance reviews would include safety criteria. Safety activities would infiltrate every aspect of the organization from production to management. Safety responsibilities would start to be shared throughout the organization. Hazard awareness would move center stage in all activities. Employees would begin to participate with greater frequency and quality than before. Management would encourage this participation. Teamwork would be the rule rather than the exception. Participation from the top of the organization to the bottom would become evident. Safety would be considered as not only a budget item, but also considered with respect to company profits, management bonuses, and operational achievement. Contractors would begin being held accountable for the same participation.

Moving along on the continuum of safety development, the next stage is the empowerment stage. At this level on the continuum employees have the responsibility for safety activities and are empowered to make decisions that affect safety. This is often without a tremendous amount of input or impedance from their superiors. The safety manager begins to be involved with others in the organization to accomplish safety-related tasks. Some would participate out of empowerment, and some out of responsibility. This stage is often bombarded with hypocrisy. Management has moved their lips with empowerment but continue to wave the stick of micromanagement, in essence trapping some of the newly empowered players into the mold of forced participation and unwanted responsibility.

At this level, progression from one stage to the next begins to show dramatic benefits. Progression from this stage would have employees making decisions previously made by management. Management would cease to issue mandates and begin to act as a facilitator and advisor. Employees would begin to create solutions on their own, thinking through projects, processing problems and circumstances. Statistical and analytical tools would all be understood and utilized by the employees themselves. Employees would make program adjustments on their own to optimize participation, awareness, and results. Programs would be held in high esteem.

In the empowerment stage, supervisors and employees would begin to plan and conduct safety meetings, perform area inspection, and watch for hazards. This first glimmer of light begins to breed ownership, the next stage on the safety system continuum. Ownership begins to trickle down to the employees, often without input or mandates from management. In the ownership stage, all levels of the organization are involved. Employees may facilitate safety meetings, conduct accident investigations, track program statistics, and perform job hazard evaluations, all with little or no participation from management. Employees at all levels are involved and proud of it. They sense the program is beneficial to them and their families. They want the program to succeed.

Next is the stage of motivation. Upon reaching this stage, benefits beyond comprehension by previous levels are reaped. At this level, systems will often begin to run themselves. In the motivation stage, employees participate because they are driven to participate by an inner need—they desire to participate. The employees recognize the benefits and want the benefits for themselves. Participation becomes fun and rewarding. Even family members add additional incentive to participate and perform safely.

Once a safety system reaches the motivation stage on the safety continuum, mandates for participation are no longer needed. Employees have long since bought into the idea of making safety an integral part of their work and their lives. Safety managers act only as facilitators for compliance issues, audits, complicated safety issues, and as liaisons with management. Management considers safety a business unit, no longer a needed dose of bad medicine. Safety is measured by the same metrics as other business sectors, such as production and maintenance.

Your organization may not fit into any single stage on the safety continuum. It may be gradually improving from one stage to the next. It may even have characteristics of several stages. Movement from one stage to the next is progressive, that is, it does not happen overnight. Although specific changes may further the progression along the continuum, it is important to remember continuum change means changing employees' values and attitudes. As in any other area of life, values and attitudes are often slow to change. Progression from one stage of the continuum to several stages further along the continuum is not impossible; however, it is difficult and not likely. Movement is most common in methodical, systematic increments.

Attainment of the motivated stage on the safety continuum is not achieved by large numbers of organizations, nor are there large numbers in the elite status of other areas of measurement such as *Fortune 500* or AAA-rated. Once the motivation stage is attained, it does not ensure an organization will maintain that status. Changes in management, company finances, personnel changes, and personalities all act as factors for fluctuation on the continuum.

An evaluation of where your organization is on the safety continuum may give understanding as to why the current level of quality and effectiveness is what it is. Purposeful progression through the safety continuum can render benefits with regard to not only accidents but financial bottom lines, employee morale and efficiency, and production costs.

In her book *Jesus CEO: Using Ancient Wisdom for Visionary Leadership*, Laurie Beth Jones talks of the need for leaders to have a plan and to work the plan. She says, "A good leader has a plan that consist of changing simple pictures. Just because a group of people has a bunch of boards, hammers, and nails does not mean that they are building a house or even anything recognizable. Sometimes leaders think they are doing their job just because there is a lot of hammering going on. As a society, we like the sound of hammering, but we are uncomfortable with the sound of thinking, which is silence."[4]

A safety program must involve employees while requiring change in the management system at the same time. One without the other is like spinning the wheels of the organization, most often to the detriment of the other. Too often today, our managers fail to see the necessity of changing the simple pictures day by day, employee by employee. Progress must be made one step at a time. Progress is a never-ending task. Thus, continuous improvement is the only way an organization's safety performance will improve. There are many different approaches to how to improve your organization's safety program. Some are good, some are bad, some will work, and some won't. However, none of them will improve safety performance without constant improvement on these fundamentals.

The ultimate goal of constant improvement must be to eliminate accidents, injuries, and illnesses from both the workplace and away from work. If the lessons of the quality movement were applied to safety and health, it would be clear to everyone that the emphasis on injury rates is bad management that leads to bad outcomes.[2]

In his book *Measuring, Managing, and Maximizing Performance*, Will Kaydos says this about improvement, "What does 'improve the process' mean? It means improving its inherent capabilities, such as efficiency, reliability, or versatility. Does giving speeches, putting up posters, or holding contests improve the process? Not really. They may be worthwhile activities, but they don't make any fundamental difference in a production system's capabilities. ...Without changing the basic activities in a production process, no lasting change in its output can be expected. ...Almost any situation can be temporarily improved by giving it more attention, but unless the fundamental capabilities of a production process are increased, the gains will not last."[3] Kaydos is talking about making fundamental change in the processes of production, but the same truths are evident for fundamental change in the safety improvement process.

The quality paradigm measures the process, not the outcome. Exposure to hazards is the process that causes injuries and illnesses. When safety is viewed in this light, performance measurement takes on a whole new look. We can easily move from trailing indicators and reactionary metrics such as lost time days and OSHA recordables to metrics that are more in step with the way other business sectors are measured.

Below are some specific safety-performance measurement criteria for a metric as simple as "safety meetings." Let's look at them and see how we would measure them in terms of constant improvement utilizing the concepts of Business Measurements for Safety Performance.

Safety Meetings
(Measurement just starting on the safety continuum)
(Ten-point maximum — 2 points each)

- 2 points if 100% of employees attended at least one safety meeting this month.
- 2 points if 100% of required training is completed.
- 2 points if all employees have been trained on all company standard operating procedures in the last 12 months.
- 2 points if all safety meetings have written documentation for who, what, where, and when the training was.
- 2 points if nonsupervisory employees presented safety meetings.

Following is a look at the same metric, safety meetings, but considered from a perspective further along the safety continuum development. Neither of these examples is intended to represent the ultimate measurement system for safety meetings. Rather, these figures demonstrate how constant improvement can be achieved by continually "raising the bar" for safety performance.

Safety Meetings
(Measurement just starting on the safety continuum)
(Ten-point maximum — 1 point each)

- 1 point if 100% of employees attended at least one safety meeting this month.
- 1 point if 100% of required training is completed.
- 1 point if all employees have been trained on all company standard operating procedures in the last 12 months.
- 1 point if all safety meetings have written documentation for who, what, where, and when the training was.
- 1 point if nonsupervisory employees presented safety meetings.
- 1 point meetings are held on a regular and consistent basis.
- 1 point if safety meetings include material other than videos.
- 1 point if employees must complete some information retention, test, or comprehension verification document.
- 1 point if the training included hands-on learning.
- 1 point if employees rather than supervisory personnel track needed training and plan training needs and applications.

Either of these examples can be altered along the way of constant improvement. For example, when a specific criterion has been met on a consistent basis, a more difficult criterion may be substituted. Certain considerations should be understood and followed when substituting criteria. Here are some examples:

- Remove or change a specific criterion only after a predetermined time.
- A criterion added because it was viewed as needing improvement should not be removed until it has been successfully completed and is considered a core competency of the organization.

- Once a criterion has been removed, close auditing of that criterion should be maintained to ensure the level of proficiency is maintained.
- When adding or removing criteria from the system, do not fractionate the weight of any criterion. In other words, maintain the consistent weighting of all criteria.
- If a criterion is especially important, instead of increasing the weighting, add criteria that will help support the achievement of the original criterion.

Tracking safety performance in this way allows an organization to examine virtually any aspect of the safety system. Once a parameter in need of constant improvement has been identified, employees can determine what objectives would be appropriate to help achieve constant improvement in that specific area. This moves an organization away from "let's improve safety" to specific achievable goals like, "let's increase the quality of safety meetings to achieve a 90% employee approval rating."

Now that we've seen how easy it is to put safety criteria into quantifiable terms, let's look more closely at how safety can be included with an organization's other business sectors.

NOTES AND REFERENCES

1. Jones, Laurie Beth, *Jesus CEO: Using Ancient Wisdom for Visionary Leadership*, Hyperion, New York, 1995, p. 88.
2. Mirer, Franklin E., Injury Rates Don't Tell the Whole Story, *Safety & Health*, August 1998, p. 62.
3. Kaydos, Will J., *Measuring, Managing, and Maximizing Performance*, Productivity Press, Cambridge, MA, 1991, p. 2.

8

Including Safety
as a Business Sector

*Hopefully, we've defeated the prospect of ever re-
turning to the simple, one-size-fits-all model.*[1]

Tracey Weiss

We have talked a lot about leading and trailing indicators. This is
not something developed just for safety performance measurement. Most of the disciplines use leading indicators as a way for
management to predict emerging problems. This is true in the
economics field with an emphasis placed on leading economic
indicators, i.e., housing starts, unemployment figures, Gross Domestic Product, Foreign Trade Deficit, and many others. We would
think it "reactionary" if the Federal Reserve waited until the country was in a severe economic depression before they opted to
adjust interest rates. We have come to expect the Federal Reserve
to monitor the leading indicators and make adjustments to improve
the economic performance of the country.

Think of the medical field. Doctors use "leading indicators" to
diagnose disease, prescribe medication, and initiate treatment plans
for patients. Imagine a doctor waiting until the patient died to say,
"I think we have a problem with the blood flow." We expect the
doctor to use the indications known to make skillful adjustments in

the treatment plan. In a business setting, we expect our chief financial officer (CFO) to monitor a multitude of indicators, both internal and external to our organization. Armed with these indicators, we then expect the CFO to make recommendations to the organization to enhance the overall performance. These same principles apply whether we're talking about the domestic economy, medical treatment, corporate stability, or safety performance.

It's all about improving the performance. By whatever system or program name or acronym, constant improvement in performance is the bottom line. "Achieving maximum performance is a balancing act, not a simple problem of optimizing one variable. Management must determine the most important factors for the entire company and assign departmental objectives and performance measures which are consistent with them."[2] In safety this means looking at leading indicators, not trailing indicators. We must measure the presence of safety, not the absence of safety. Unsafe behaviors are typically ignored until they become accidents, and then they become unacceptable. With the technology of the new millennium, there will be no excuse for limiting the effort to "measure what you manage" to the hard numbers.[3] "Hard numbers" will always be around. They are how we make comparisons, how ranking systems pick the "top ten," how a multitude of decisions are made—we even talk about them over dinner. But, beyond all that, one must look deeper into the "nitty-gritty" of performance measurements. In all business sectors that is done by managing leading indicators that are known to affect the trailing indicators.

The case for a business sector measuring performance by many indicators is an easy one. Just look anywhere, at any business sector, no matter how large or small and you will find this method of performance measurement in place. We have moved a long way from the notion that there is nothing new in performance management. Hopefully, we have eliminated the prospect of ever returning to the simple, one-size-fits-all model.[4] The hard sell is determining why safety performance is not typically included in this type of performance-based measurement. Just as management is responsible and held accountable for all other business sectors it must be held accountable for the safety and health of its employees. Until the accountability of safety is aligned with other business sectors there will be no acceptance of the responsibility for safety, and thus no significant improvement.

An appropriate way to begin tying safety to business may be to list some of the areas where it obviously makes sense to consider safety in a business light rather than in some other, less important area.

Safety is Business

- Worker compensation costs directly effect the bottom-line profit of the company.
- Insurance premiums are driven higher by poor safety performance, driving the costs of doing business higher.
- The costs associated with injured employees' health care continue to skyrocket with other medical costs.
- Injured employees cause morale to be low and decrease productivity.
- Injured employees must be replaced with additional man-power, increasing overtime and training costs of replacement manpower.
- Quality decreases due to inexperienced workers filling in for injured employees.
- Repair costs for equipment damaged in accidents can be high.
- Lost production time due to damaged equipment is irreplaceable.
- Accident investigation time for everyone concerned equates to lost productivity.
- Legal liabilities can amount to millions of dollars in settlements, fines, and legal representation.

This is only the tip of the iceberg, so to speak. The reasons are as numerous and valid as in any other business sector. Once put in this light, modern management can make no choice but to quantitatively measure safety in the same ways and with the same importance as in any other business sector. You will notice that nowhere have we said, "safety is number-one" or "safety must come first." Those old adages are part of the problem the safety industry must undo to put safety on the same ground as production and other business sectors. They are simply not true and only boil the pot of hypocrisy. If you think safety must be "first," then you're agreeing to continually beat your head against the proverbial wall. The pressure for profit, a competitive marketplace, and a hundred other factions struggling for prominence will continually try to unseat the "safety-first" mentality and replace it with "production," "profit," "quality," or whatever else seems important at the time.

Safety must become an integral part of everything we do in industry; not before or above quality, but intertwined with quality; not before production, but woven into the production process so closely that there is no separation between the two. It is not a choice between safety and profit; it is a marriage between them that must be strengthened at every occasion.

The standards that are used need not be concretely set and forevermore held to be true. I'm not advocating changing the measurement tool at every whim, but to step back and see if how we are measuring safety is the way that makes the most sense and benefits our organization the most. We must decide not only what should be measured, but how it should be measured. "It should be noted that all of the international standards are purely arbitrary and have become accepted by custom and law. At one time the standard for the yard was from the tip of the king's nose to the end of his outstretched hand. That supposedly worked quite well until the king died and was replaced by Shorty the Ninth, causing fortunes to be made and lost overnight. Eventually the English and metric systems of standards were developed to provide the stability necessary for trade. They are equally valid standards, although only a few backward countries like the United States still use the English system."[5] The point being, the metric currently being used to measure safety performance in your organization may be in need of an objective review.

We end this chapter without citing what other specific business metrics should be duplicated for the measurement of safety performance. It would be impossible for me to list what your organization uses to measure, say, operations or shipments. Every organization already has its own culture and its own performance measurements in place. Hopefully, you can now determine that the primary concern is measuring safety by the same methods and importance that your organization measures other business sectors and, secondly, that safety be measured with a multitude of both leading and trailing indicators.

While in this chapter I want only to bring a comparison and rationale for measuring safety performance as a business sector, in Chapter 10 we will discuss many specific possibilities for safety-performance measurement.

NOTES AND REFERENCES

1. Weiss, Tracey B. and Hartle, Franklin, *Reengineering Performance Management Breakthroughs in Achieving Strategy through People*, St. Lucie Press, Boca Raton, FL, 1997, p. 103.
2. Kaydos, Will J., *Measuring, Managing, and Maximizing Performance*, Productivity Press, Cambridge, MA, 1991, p. 54.
3. Weiss, Tracey B. and Hartle, Franklin, *Reengineering Performance Management Breakthroughs in Achieving Strategy through People*, St. Lucie Press, Boca Raton, FL, 1997, p. 195.
4. Weiss, Tracey B. and Hartle, Franklin, *Reengineering Performance Management Breakthroughs in Achieving Strategy through People*, St. Lucie Press, Boca Raton, FL, 1997, p. 103.
5. Kaydos, Will J., *Measuring, Managing, and Maximizing Performance*, Productivity Press, Cambridge, MA, 1991, p. 16.

9

Getting Buy-In
from the Troops

*Doing everything the same old way is sure to pro-
duce the same old results.*[1]

Will J. Kaydos

WILLINGNESS TO CHANGE

If you truly want to improve the performance of your business or
department, you must be willing to change anything and everything. Throw out all the rules and challenge every assumption.
Consider everything a variable. Tear the system apart and start
over. Get someone who doesn't know "every widget maker does
it that way" to review you situation.[2] This is a prerequisite to
getting everyone on board. Great safety initiatives fall like lead
balloons without support from all levels of an organization. Using
the same mindset that provided poor safety performance will only
replicate poor safety performance.

This chapter will discuss buy-in from management, and buy-in
from employees and supervisors. Keep in mind that no one level
of an organization's buy-in, no matter how committed, can work
unless there is buy-in from all levels. Most will agree, though, that
commitment from top management is the first step toward buy-in
elsewhere in the organization. Part of getting buy-in is selling
something that's worth buying. It must fit the culture, style, and
character of the organization.

TRUE COMMITMENT FROM TOP MANAGEMENT:

A growing recognition is emerging among safety management that the meaning of the word "commitment" is "no injury is acceptable, all injuries are preventable, and we will do whatever it takes to prevent injury." This position puts all employees—the CEO, management, and workers—on the same footing. Such a definition is commonly found among those who embrace the safety philosophy now commonly known as the "zero injury concept."

Emmitt Nelsen, in his article in *Professional Safety,* describes how buy-in and commitment must come from top management before it can be bought-into by workers:

> The Construction Industry Institute (CII), Austin, Texas, provided the answer in 1993 after completing a $1 million research project. The project was aimed at determining how some employers were able to work millions of hours without serious injury to their employees while others found it difficult to report fewer than six (the national average) lost workday cases per 200,000 hours worked.

> Research Results

> The key finding: In the zero lost workday companies, the CEO always placed a key safety expectation before management. It was (paraphrased) as follows:

> We will do our work without an injury. It is my (the CEO's) belief that all injury can be prevented. It is my expectation that there be no worker injury on our projects. And if an injury does occur, it will not be viewed by me as acceptable performance! I personally will be involved in determining how management failed. We will not set goals for injury! Our commitment is to ZERO injury! This is not a statistical management effort. It is a complete devotion to the elimination of unsafe behavior by all employees—management and workers alike.[3]

"I will be personally involved." Those are strong words compared to most managers' commitment to safety. That kind of commitment is an essential link in changing the culture of an organization. "A leader who is not passionately committed to the cause will not draw much commitment from others. The world will make way for someone who knows what he or she wants, because there is not much competition when it comes to passionate commitment. The Scriptures include examples of God's reaction to non-commitment. Jesus said, 'It is better to be hot as fire or cold as ice, because if you are lukewarm I will spit you out.' (Rev. 3:15) He constantly warned his staff of the importance of commitment to his cause. He was willing to walk the road alone—and did—when it came to the price of his commitment."[4]

Some believe that traditional injury frequency rates have no value because they rely on reactive, historical measures rather than proactive approaches to safety. Another criticism is that these statistics encourage management to react without providing accurate feedback on the effectiveness of the safety effort. Other experts believe that employee behavior is the most valuable indicator of safety performance. They say that if organizations can identify safety-related behaviors, they will have the only tool needed to improve workplace safety. Despite their differences of opinion on a standard of measurement (metric), most experts agree that management commitment is the principal determinant of safety performance. The bottom line is that all of these measures are important.[5]

EMPOWER THE PEOPLE

Leaders must share information and the authority that goes with it. This way they can empower others to do the right thing in ways that will offer fulfillment, not only on an individual level, but on a global basis as well. To grant authority is to leverage one's gifts.[6] A manager's responsibility is to find the ways that new initiatives will work, not to dwell on the hundreds of ways they will not. In most cases, the manager is not the link in the organization that has the answers. A manager's place in the organization is to create an atmosphere, a culture where the employees who do have the answers can work optimally. Managers forget they are managers. They seem to have an unavoidable tendency to be doers, not managers. Managers must become facilitators, not hindrances to accomplishing anything meaningful. This is not a new problem. Wellington knew all too well how detrimental micromanagement was.

This note was received by the British War Office in London in 1812:

Gentlemen:

Whilst marching from Portugal to a position which commands the approach to Madrid and the French forces there, my officers have been diligently complying with your request...

We have enumerated our saddles, bridles, tents poles, and all matter of sundry items for which His Majesty's Government holds me accountable. I have dispatched reports on the character, wit, and spleen of every officer. Each item and every farthing has been accounted for with two regrettable exceptions...

Unfortunately the sum of one shilling and nine pence remains unaccounted for in one infantry battalion's petty cash, and there has been a hideous confusion as to the number of jars of raspberry jam issued to one cavalry regiment during a sandstorm in Western Spain...

This brings me to my present purpose, which is to request elucidation of my instructions from His Majesty's Government, so that I may better understand why I am dragging an army over these barren plains. I construe that perforce it must be one of two alternative duties, as given below. I shall pursue either one to the best of my ability, but I cannot do both.

1. *To train an army of uniformed British clerks in Spain for the benefit of the accountants and copy-boys in London, or perchance,*
2. *To see to it that the forces of Napoleon are driven out of Spain.*

Your most obedient servant,
Wellington

Humorous as this letter may seem, Wellington was experiencing what today's employees call "micromanagement." One sure way for your safety program NOT to work is to fail to empower your employees. The culture in which management empowers its employees is one in which trust and commitment flourish. A culture where micromanagement is king only frustrates everyone involved and continues to get the same old results, typically poor safety performance.

If you as a leader or manager intend to accomplish anything significant, the first step toward attaining your goal is to create a team. Yet, many people still feel they must do everything alone. We still have John Waynes and Super Moms who think it is wrong or a sign of weakness to ask for help.[7] Delegation of authority requires a tremendous amount of trust. Perhaps there are so many confused employees because there are so many fearful people at the top. If leaders operate out of fear, they cannot delegate. A leader who does not delegate will end up with a group of "yes" people who will ultimately lead to his or her demise.

DECIDE WHAT YOU WANT

Management leadership in safety and health goes beyond publishing a policy and establishing goals. All levels of management must be convinced that safety increases employee productivity, reduces losses, and is an important organizational priority. Members of management must then frequently communicate this belief and vision by word and example. Management also needs to let supervisors and employees know that they must actively fulfill responsibilities contributing to specific safety goals. Upper-level managers must also be personally involved in planning, reviewing, and recognizing activities to maintain the safety effort's internal momentum. Through involvement, management shows its commitment and demonstrates the importance of the safety program to all employees.[8]

DuPont's Beaumont, Texas, plant safety performance was good in 1990. The employees wanted to make it better. They chose 170 criteria they felt were important to an overall safety program. This number is similar to the number of criteria described in

Chapter 11. The plant decided that performance indexing was the best way to address the areas of the safety program that needed improving or needed continuous monitoring. Putting the indexing plan in its simplest form, it had four main points: (1) conduct in-depth analysis of site safety problems, (2) select between five and ten metrics that relate to the specific safety problem already identified, (3) measure and plot the site's performance regularly and communicate the results, and (4) actively involve employees in these processes.[9]

How did performance indexing affect Beaumont's safety record? After three years of performance indexing, Beaumont had boosted its performance to fourth place among DuPonts largest sites, had not lost work days in three years, and had a total OSHA recordable rate of only 0.28.

What DuPont is calling performance indexing is nothing more than Business Measurements for Safety Performance; basically, weighing a multitude of performance measurements in order to gauge safety performance.

CREATE A CULTURE THAT CARES ABOUT PEOPLE

"Safety Culture" can be defined as a set of shared attitudes, values, goals, and practices that characterizes a company's safety effort. In most cases, the parts that make up the safety effort develop through years of individuals working with the same set of coworkers who follow safety policies and procedures that have been in their company for years. These policies are seen largely as ways to appease OSHA. Culture is different, it is beliefs and values. "Culture is the way things are around here."[10]

Therefore, changing an existing safety culture can be a very difficult thing to do. It shakes up employees' routines. Unfortunately, cultural change is almost always necessary when implementing a behavior-based safety process in an organization.[11] In fact, a cultural change is required with implementation of any new idea that does not fit with "the way we've always done it" method of thinking.

Once the culture begins to change at the top, it does not take long for the change to filter down through the organization. Cultural change is a long-term process. Don't expect one meeting, memo, or announcement to magically change the culture. Looking for a dramatic breakthrough is only counterproductive; slow but steady change is an approach that is more realistic.

The measures used must instill a belief that they are fair, equal, and beneficial. They must focus on the same things as those in other sectors that are perceived to have worth to the organization.

SURE FAILURE

The following obstacles are often stumbling blocks for organizations attempting to implement behavior-based safety cultures:

1. Supervisors are widely viewed as "safety police," thus, safety is seen as negative.

2. Supervisors and management don't practice what they expect workers to practice.

3. There is lack of trust among workers.

4. Safety efforts are dependent on the occurrences of injuries. At-risk behavior is overlooked until someone gets injured.

5. There is an overemphasis on production or quality. Safety is compromised to speed up production or to enhance quality.

6. The company doesn't recognize that its "safety culture" is made up of more than what may be visible, such as policies, procedures, and behaviors.[12]

Rhoden's list is a good start for understanding roadblocks to cultural change. Other roadblocks that exist in your organization may need to be examined more closely. In fact, if these obstacles exist, you should proceed with caution as you embrace cultural change.

Roadblocks to Cultural Change

- Downsizing or rightsizing is or has just taken place
- Union negotiations have been hard
- Heavy use of contractors exists for problem jobs
- Management has a heavy hand
- Little empowerment exists
- Management / employee relationships are tense
- Management is playing games with numbers
- Fault-finding is typical, looking for blame is common
- Information is not shared readily with employees

Just as there are some cultural roadblocks, there are some "must-do" factors that need to be included before any successful cultural change can begin. Some of these essential elements are presented in the following list. Again, this is only a fraction of the "atmosphere" that should exist for successful cultural change

Essentials for Cultural Change

- Performance targets must be well-defined and clearly communicated to everyone
- Top management and executives must spend time and be actively involved
- There must be a recognition by everyone that a change needs to occur
- All levels of the organization must be trained and educated on the areas to be measured and/or changed
- All employees must be involved and participate
- The "plan" must be communicated to everyone, understood by everyone, practiced by everyone, and encouraged by everyone.
- An accurate assessment of the current culture must be made and communicated to all employees
- All levels of the organization must have meaningful, active roles in establishing the goals for cultural change
- Employee participation at every level is absolutely essential

Hopefully, this chapter has given some useful guidelines as to how to get everyone from front-line employees to the CFO involved in making a cultural change in your organization. Changing from traditional trailing indicators to a broad-based, leading indicator safety performance measurement system is a cultural change, to say the least. The following chapters will give more details on how to structure an aggressive safety performance measurement system.

NOTES AND REFERENCES

1. Kaydos, Will J., *Measuring, Managing, and Maximizing Performance*, Productivity Press, Cambridge, MA, 1991, p. 4.
2. Kaydos, Will J., *Measuring, Managing, and Maximizing Performance*, Productivity Press, Cambridge, MA, 1991, p. 4.
3. Nelsen, Emmitt J., Safety Commitment Redefined, *Professional Safety*, December 1998, p. 41–43.
4. Jones, Laurie Beth, *Jesus CEO: Using Ancient Wisdom for Visionary Leadership*, Hyperion, New York, 1995, p. 51.
5. Herbert, David A., How to measure where your organization has been, where it's at and where it's going, *OH&S Canada*, March/April 1995, p. 54–60.
6. Jones, Laurie Beth, *Jesus CEO: Using Ancient Wisdom for Visionary Leadership*, Hyperion, New York, 1995, p. 264.
7. Jones, Laurie Beth, *Jesus CEO: Using Ancient Wisdom for Visionary Leadership*, Hyperion, New York, 1995, p. 91.
8. Kolosh, Kenneth P., *Sizing-up Safety*, National Council on Compensation Insurance, Boca Raton, FL, 1998, p. 56–59.
9. Herbert, David A., How to measure where your organization has been, where it's at, and where it's going, *OH&S Canada*, March/April 1995, p. 54–60.
10. Petersen, Dan, *Analyzing Safety System Effectiveness*, 3rd edition, Van Nostrand Reinhold, New York, 1996, p. 65.
11. Rhoden, Travis, Changing Safety Culture: A Common Obstacle in Implementing Behavior-Based Safety, *Industrial Hygiene News*, September 1998, p. 42.
12. Rhoden, Travis, Changing Safety Culture: A Common Obstacle in Implementing Behavior-Based Safety, *Industrial Hygiene News*, September 1998, p. 42.

10

Tracking Performance

If you want to manage performance, you must measure performance.

Daniel Patrick O'Brien

This chapter will present several valid approaches to changing the way safety performance is measured. All of them agree in principle that safety performance measurement is not the sole output of trailing indicators such as the number of lost time accidents or the organization's OSHA recordable incidence rate. Each of the systems or approaches comes from experts in the safety and health field; some experts come from the practitioner's side, others from academia, and still others from consulting. Each of them brings years of experience in the business of preventing injuries and accidents. As I've said before, they don't always agree. That's not the important part, the important part is that you are able to glean an idea or concept that will work for your organization and utilize that concept for the reduction of injuries and illnesses.

Dr. Renis Likert developed a survey that measured employee perception in ten areas. This list is presented early on because, in most cases, a perception survey is a good place to start. If an organization pulls a great new program out of its hip pocket and begins to implement it, it will most likely be met with failure. The employees must be involved from the very beginning of any new endeavor initiated by an organization. The employee's initial involvement should come from giving his or her input about how things are currently. Without a good, clear understanding of where things stand now, it makes it very difficult for an organization to decide where it wants to go.

Likert's 10 Areas of Employee Perception Evaluation

1. Confidence and trust in the organization.
2. Interest in the subordinate's future.
3. Understanding of, and the desire to help overcome problems.
4. Training the subordinate to improve.
5. Teaching the subordinate how to solve problems rather than finding answers.
6. Giving support by making available required physical resources.
7. Communicating information that the subordinate must know to perform the job, as well as information needed to identify more with the operation.
8. Seeking out and attempting to use ideas and opinions.
9. Approachability of supervisors, managers, and executives.
10. Giving credit and recognizing the subordinate's accomplishments.

Employee involvement improves the quality of decisions by improving the quality and quantity of information available to decision makers throughout the company. Since the person doing the work knows more about his or her problems than anyone else does, every employee is a source of valuable and unique information.[1] Knowing that before you decide where to go you must first determine where you are, the employees are the most logical ones to begin asking questions. Employee perception surveys are among the most underused tools the modern organization has. High-quality perception surveys are on the market that are statistically validated and provide an extremely accurate picture of what your current organization looks like from people in the organization's perspective. Short of canned, statistically validated surveys, an organization can quickly and easily accumulate useful information from its employees that can prove invaluable in directing future monetary funds, manpower, and effort expended in the attempt to make significant change in safety-performance measurement. Perception surveys should be used as diagnostic tools. What's the climate like? What's the safety culture (Safety and corporate)? They provide feedback that can be used by employees as well as supervisors. Survey information should be collected from all levels in the company. The data collected must be conveyed to all employees before it is truly helpful and worth collecting in the first place.

In his book, *Analyzing Safety System Effectiveness*, Petersen sets out 21 areas in which he believes world-class companies tend to be effective. He notes that these do not comprise a canned program for safety excellence, but are 21 areas worthy of your company examining with regard to its performance. Petersen's 21 points to analyze follow.[2]

Petersen's 21 Points of Effectiveness

The management system you have in place to achieve continuous improvement of your safety process.
• Your accident investigation process.
• Whether your employees are involved.
• Your operating practices.
• Your discipline policies.

The management system you have in place to build a positive safety culture.
• What climate actually exists.
• Management's real credibility.
• Support for safety.
• The recognition the worker receives (or does not receive) for contributions.
• The general attitudes of your employees.
• The amount of stress that people feel daily on the job.

The management system you have in place to improve the skills of your supervisors and managers.
• Effectiveness of your supervisory training.
• The perception of the quality of your supervisors.
• Your goal-setting process.

The management system you have in place to improve the skills of your employees.
• Their training.
• Your handling of the new worker.
• Your communication effectiveness.

The management system you have in place to improve employee behavior.
• Your safety contacts.
• Your alcohol and drug programs.
• Your awareness programs.

The management system you have in place to improve physical conditions.
• Your inspections.
• Your hazard correction procedure.

You can see the case for safety performance measurement is a strong one. It really is not an issue of whether safety performance measurement should be measured by a multitude of metrics, but how many and which metrics will most closely align with the culture within your organization and achieve constant improvement.

In today's safety world we hear a lot about Behavior-Based Safety. I have not specifically addressed Behavior-Based Safety in this text, but not because I disagree or don't feel that it is viable. Behavior-Based Safety is at the core of Business Measurements for Safety Performance, which revolves around the concept of moving away from "trailing indicators" (results) and moving toward "leading indicators" (behaviors and activities). So while Behavior-Based Safety and Business Measurements for Safety go hand-in-hand, Behavior-Based Safety is but one facet of the approach of Business Measurements for Safety Performance.

In his book, *The Psychology of Safety—How to Improve Behaviors and Attitudes on the Job*, Scott Geller sums up his book by offering 50 principles of a Total Safety Culture. We've seen that most safety and health organizations—nationally-known safety consultants, speakers, authors, and safety "wannabees," myself included in one of those descriptions—have their lists of "the ten most," "the three essential," the collective (fill in your own number) of special points, principles, or fundamental have-to's for safety-performance excellence.[3] I think Geller's principles are worthy of listing here. Although some may be difficult to fully understand without referring to his actual text, I believe by listing them here, the principles may help the reader to grasp yet one more philosophy of how to achieve safety excellence.

Geller's list is comprehensive and unmistakably important to the overall safety culture, safety performance, and constant improvement of the safety system in an organization. Where Business Measurements for Safety Performance comes from is that Geller's points are important, just as Petersen's and Likert's points are important. I contend that you need not choose, rather pick and choose what works best for your organization. In some organizations Geller's 50 points could fit in nicely and reap tremendous safety-performance improvement. In other organizations, some just as good, anybody's 50 points for anything are going to go over like lead balloons. Looking back to Chapter 7 and the safety continuum, each organization must evaluate where it is on the safety continuum and then decide what areas are in need of improvement. We'll see in Chapter 11 how even 50 points can become small when a total safety system is evaluated at every level.

So what would my "critical points" look like? My critical points for constant improvement of safety performance would be pretty generic. My reasoning for that is that you can see by the information presented in this text that there are literally hundreds of valid areas of concern, some more critical than others, but different ones critical to different organizations. With this text and many other available texts, many of them mentioned in this book's references, are available to help home in on exactly what you need to measure for your organization.

Scott Geller's 50 Key Principles of a Total Safety Culture

1. Safety should be internally — not externally — driven

2. Culture change requires people to understand the principles and how to use them.

3. Champions of a Total Safety Culture will emanate from those who teach the principles and procedures.

4. Leadership can be developed by teaching and demonstrating the characteristics of effective leaders.

5. Focus recognition, education, and training on people reluctantly willing, rather than on those resisting.

6. Giving people opportunities for choice can increase commitment, ownership, and involvement.

7. A Total Safety Culture requires continuous attention to factors in three domains: environment, behavior, and personal.

8. Don't count on common sense for safety improvement.

9. Safety incentive programs should focus on the process rather than outcomes.

10. Safety should not be considered a priority, rather a value with no compromise.

11. Safety is a continuous fight with human nature.

12. Behavior is learned from three basic procedures: classical conditioning, operant conditioning, and observational learning.

13. People view behavior as correct and appropriate to the degree they see others doing it.

14. People will blindly follow authority, even when the mandate runs counter to good judgment and social responsibility.

15. Social loafing can be prevented by increasing personal responsibility, individual accountability, group cohesion, and interdependence.

16. On-the-job observation and interpersonal feedback is key to achieving a Total Safety Culture.

17. Behavior-based safety is a continuous DO IT process.

18. Behavior is directed by activators and motivated by consequences.

19. Intervention impact is influenced by: amount of response information, participation and social support, and external consequences.

20. Extra and external consequences should not overjustify the target behavior.

21. People are motivated to maximize positive consequences (rewards) and minimize negative consequences (cost).

22. Behavior is motivated by six types of consequences: positive vs. negative, natural vs. extra, and internal vs. external.

23. Negative consequences have four undesirable side effects: escape, aggression, apathy, and countercontrol.

24. Natural variation in behavior can lead to a belief that negative consequences have more impact than positive consequences.

25. Long-term behavior change requires people to change "inside" as well as "outside."

26. All perception is biased and reflects personal history, prejudices, motives, and expectations.

27. Perceived risk is lowered when a hazard is perceived as familiar, understood, controllable, and preventable.

28. The slogan "all injuries are preventable" is false and reduces perceived risk.

29. People compensate for increases in perceived safety by taking more risks.

30. When people evaluate others they focus on internal factors; when evaluating personal performance, they focus on external factors.

31. When succeeding, people overattribute internal factors; but when falling, people overattribute external factors.

32. People feel more personal control when working to achieve success than when working to avoid failure.

33. Stressors lead to positive stress or negative distress, depending on appraisal of personal control.

34. In a Total Safety Culture everyone goes beyond the call for the safety of themselves and others — they actively care.

35. Actively caring should be planned and purposeful, and focus on environment, person, or behavior.

36. Direct, behavior-focused activity caring is proactive and most challenging, and requires effective communication skills.

37. Safety coaching that starts with Caring and involves Observing, Analyzing, and Communicating, leads to Helping.

38. Actively caring can be increased indirectly with procedures that enhance self-esteem, belongingness, and empowerment.

39. Empowerment is facilitated with increases in self-efficacy, personal control, and optimism.

40. When people feel empowered, their safe behavior spreads to other situations and behaviors.

41. Actively caring can be increased directly by educating people about factors contributing to bystander apathy.

42. As the number of observers of a crisis increases, the probability of helping decreases.

43. Actively caring behavior is facilitated when appreciated, and inhibited when unappreciated.

44. A positive reaction to actively caring can increase self-esteem, empowerment, and belongingness.

45. The universal norms of consistency and reciprocity motivate everyday behaviors, including actively caring.

46. Once people make a commitment, they encounter internal and external pressures to think and act consistently with their position.

47. The consistency norm is responsible for the impact of "foot-in-the-door" and "throwing a curve."

48. The reciprocity norm is responsible for the impact of the door-in-the-face technique.

49. Numbers from program evaluations should be meaningful to all participants and direct and motivate intervention improvement.

50. Statistical analysis often adds confusion and misunderstanding to evaluation results, thereby reducing social validity.

In all truthfulness, some of the "critical points" are just a matter of personal belief or preference. A lot depends on what your organization needs and what the culture is, how dynamic the players are, and how much commitment is realistic. So my "critical points" are geared to keep everyone in the game, not ruling out those who would prefer all behavior-based approaches, or those who would align to an all-management approach. The following are Dan O'Brien's Critical Points for Safety Performance Constant Improvement.

Dan O'Brien's Critical Points for Constant Improvement of Safety Performance

- Start out by conducting an employee perception survey. It should be anonymous and from every employee in the company.
- Involve all levels of employees in every step of the process. No exceptions!
- Put real safety objectives and goals (with meat) in every employee's performance objectives.
- Make every employee responsible for doing predetermined safety activities every week.
- Eliminate micromanagement at every level of the organization.
- Empower employees with true power and authority.
- Include trailing indicators only as one piece of the safety system, no more.
- Form expert accident investigation teams.
- Use only quality material for safety meetings.
- One-on-one observations are of paramount importance.
- Safety must be integrated into every aspect of the organization. No exceptions!
- Employees, supervisors, managers, and executives all need training on how to constantly improve the safety culture.
- Place the same importance on safety as on other business sectors.
- Track behaviors and activities in a positive, encouraging way.
- Make no promises or statements of commitment that you don't intend to back up with the resources of the entire company.

I have also included a short list of the main sectors from the system above showing the progression from one year to the next. This is how constant improvement can be achieved by occasionally adjusting the criteria to push safety performance to new levels and to address areas not previously addressed.

Now, are my points any better than any of the others that I have pointed out? Not at all. They are but one more guide to help you establish a constantly improving safety culture. Following is an example of what a cursory Business Measurements for Safety Performance might look like. Again, this is only an example and may fit nicely into your organization as is, or it may need significant adjustments to be of any value at all.

Business Measurements for Safety for 1999

_____ OSHA Recordable Incidence (1998)

_____ Safety Meetings (1998)

_____ Safety Suggestions (1998)

_____ Safety Auditing (1998)

_____ Contractor Interface (1998)

_____ Housekeeping (1998)

_____ Ergonomic Efforts (New)

_____ Industrial Hygiene Issues (New)

_____ Documentation (1998)

_____ Management Involvement (New)

Chapter 11 will show how all of these methods might flow together for safety performance measurement in the future.

Business Measurements for Safety Performance

____ **Plant Security** (10 points) (2.5 points each)

- Are the plant gates tied into the plant computer system and monitored during off hours?
- Are the plant gates secured between 6:00 PM and 6:00 AM?
- Are the gates secured on weekends?
- Are the gates monitored in the control room by camera?

____ **Safety Meetings** (10 points)

- Each employee must meet his or her location's minimum requirement for Safety Meetings in order for the location to qualify for these 5 points.

- Each employee must have received training on all Standard Operating Procedures within the last 12 months in order for the location to qualify for these 5 points (90-day leeway for new employees).

____ **Safety Suggestions** (10 points)

- Each employee must meet his or her location's minimum requirement for Safety Suggestions in order for the location to qualify for these 5 points.

- Each employee must have received hazard identification training within the last 12 months in order for the location to qualify for these 5 points (All employees are exempt from this in January 1999).

____ **Safety Auditing** (10 points)

- Each location must have two interplant safety audits per year. (One each six-month period) plus one formal interplant safety audit per month in order for the plant to qualify for this 5 points.

- All findings from the last safety audit must either be corrected and signed off on, or a written corrective action plan implemented for any audit item over 30-days old for the location to qualify for this 5 points.

____ **Housekeeping Audit** (10 points)

- A housekeeping audit must be conducted in each department, area, or team that meets the minimum requirement for that location to qualify for this 5 points.

- A written monthly report must be distributed to the employees of each department, area, or team for the location to qualify for this 5 points.

Business Measurements for Safety Performance (continued)

____ **Ergonomic Efforts** (10 points)

Any two of the following ergonomic efforts must have been completed during the month for the location to qualify for this 10 points.

- A locationwide lighting survey
- A locationwide chair inspection and replacement survey
- A locationwide workstation evaluation survey
- A detailed ergonomic evaluation of a particular workstation as per SSOP 22.710
- Automation of a particular job function.
- All employees have received workstation exercise training
- All employees have received training on a back exercise program
- Pre-shift exercises are conducted at least three days per week
- Locationwide ergonomic training has been conducted in the last month
- A locationwide employee lifting survey has been completed in the last month
- A locationwide ergonomic hand tool survey has been conducted in the last month.
- Ergonomic posters are displayed in the location and changed monthly.
- Ergonomic workstation aids such as ergo keyboards and lumbar supports have been offered to all employees.
- The ergonomic SSOP has been trained on in the last month.
- An ergonomic "body check survey" form has been conducted in the last month.

(Items can only be used once per year. Other ergonomic projects may qualify for 5 points.)

____ **Industrial Hygiene Issues** (10 points)

Any two of the following industrial hygiene efforts must have been completed during the month for the location to qualify for this 5 points.

- A locationwide product leak survey conducted every day this month.

Business Measurements for Safety Performance (continued)

____ **Industrial Hygiene Issues** (10 points) (continued)

- A locationwide noise survey conducted this month.
- A locationwide hearing / noise survey conducted this month.
- A particulate monitoring survey of a high-level exposure group conducted this month.
- An exposure monitoring study conducted on any particular job conducted this month (HCN, H2S, CO, etc.).
- A comprehensive feedstock exposure study has been conducted this month.
- A comprehensive welding fume exposure study has been conducted in the shop.
- An indoor air quality study has been conducted this month.
- A vent hood inspection program was implemented this month.
- A solvent exposure monitoring study was completed this month.

A Job Safety Analysis schedule has been developed and implemented to review every task on every job description in 1999 to qualify for these 5 points.

(Items can only be used once per year. Other Industrial Hygiene projects may qualify for 5 points.)

____ **Documentation** (10 points)

- All training must be logged, both on a location-training matrix and in the Policy Controller's policy book for the location to qualify for these 5 points.

- Every SSOP must have been audited and a written report filed in the Policy Controller's policy book for the location to qualify for these 5 points.

____ **Management Involvement** (10 points)

Any two of the following Management efforts must have been completed during the month for the location to qualify for this 10 points.

- A member of management outside the plant has toured the location this month.
- The location manager has toured the entire location at least once each week this month.

Business Measurements for Safety Performance (continued)

Management Involvement (10 points) (continued)

- Every member of management has specific safety objectives listed on his or her job performance evaluation.
- Location management has reviewed safety meeting feedback and responded to employees.
- The location manager has ensured every department has met the location requirement for safety meetings this month.
- Location managers held a monthly safety meeting this month.
- All managers are distributed a copy of every accident.
- Management has established a system for addressing and responding to every safety meeting.

(Other items showing management's commitment may qualify for 5 points. Items can be used only twice per year.)

_____ **Routine Inspections** (10 points) (1 point each)

- Have fire extinguishers been checked, serviced as needed, and documented?
- Have eyewash and shower stations been tested, repaired as needed, and documented?
- Have the first-aid kits been inspected, restocked, and needed service or repairs made?
- Have all escape masks for special chemical exposure tasks been inspected?
- Has the plant evacuation siren (Fire and Tornado) been tested?
- Have all confined space rescue kits been serviced, restocked, and cleaned?
- Has the plant's weather radio been checked for availability, batteries, channels?
- Is the plant "take cover" area clean, stocked, designated, and accessible?
- Does the location have at least one stocked "EMT" responder bag?
- Have four weekly major equipment inspections been completed, documented, and repairs made?

_____ **Total Score** (maximum score 100 points)

NOTES AND REFERENCES

1. Kaydos, Will J., *Measuring, Managing, and Maximizing Performance*, Productivity Press, Cambridge, MA, 1991, p. 153.
2. Petersen, Dan, What Measures Should We Use, and Why? Measuring Safety System Effectiveness, *Professional Safety*, October 1998, p. 37–40.
3. Geller, Scott, *The Psychology of Safety—How to Improve Behaviors and Attitudes on the Job*, Chilton Book Company, Radnor, PA, 1996, p. 363–380.

11

Performance Measurement in the Future

Tomorrow's corporate heroes won't dominate or brainwash; they will instill a feeling of power in others. They will be role models. They will achieve their strategy through other people, not despite them.[1]

Tracey Weiss

ASSESSMENT TOOLS

Total Quality Management techniques in the 1980s brought about more reliable, sensitive, and responsive measurement techniques. These techniques demonstrated that performance can be used not only to track performance, but to drive performance improvements. In 1998, the *American Industrial Hygiene Association Journal* published what the author believes to be a landmark article regarding what future organizations will measure in the area of Occupational Health and Safety. Charles Redinger and Steve Levine wrote the article based on a research project from the Environmental Management Institute at the World Health Organization Collaborating Center, and the School of Public Health, The University of Michigan at Ann Arbor.[2]

This assessment instrument is sure to influence the direction and fundamental approach taken toward system management in safety and health. This assessment instrument is not presented here as the ultimate safety metric of all safety metrics. Rather, it is presented as a pattern or framework around which to base individual organization performance evaluations. Remember, what works for one safety system or culture may not work for the next, and vice-versa. Even with the ultimate metric in hand, paramount consideration must be given to what works in a given culture.

This "Assessment Instrument" (in the author's words, "list of metrics") was developed by looking at a wide range of Occupational Health and Safety Management Systems (OHSMS). In fact, more than 13 OHSMS were reviewed. Data taken were merged and then organized according to common criteria. The common criteria were reviewed in-depth to determine which OHSMS best represented these criteria, which were the most thorough, and which were on the cutting edge of industrial use and need. Four models emerged as the basis for the Michigan OHSMS Assessment Instrument (MAI). The four models were the OSHA Voluntary Protection Program; the British Standards Institute's OHSMS, BS8800:1996; the American Industrial Hygiene Association's OHSMS; and the International Organization for Standardization's (ISO) EMS ISO 14001:1996 and ISO 9001.[3]

If this sounds like a monumental endeavor, it was. The study ultimately revealed 27 separate sections, 118 OHSMS principles, and 486 measurement criteria. The MAI provides a framework that can contribute to the identification and measurement of occupational safety and health performance measurement — the ultimate list for Business Measurements for Safety Performance.

Let's stop here and say that I'm not advocating your organization copy from the 486 criteria and start a monumental, endless checklist. I present this model because it represents the future of performance measurements. In essence, all the powers-that-be should sit down, make the ultimate of ultimate list of what is important to measure, try to put it in some meaningful format, and then, there it is. This may be tremendously oversimplified, but it's the basic concept of Business Measurements for Safety Performance: find out what's important and measure it.

The study indicated that many organizations were already using many of the criteria specified in the MAI. Throughout the world, governments are seeking new OHSMSs. These countries include the United States, the United Kingdom, India, Brazil, Japan, and Australia. With the globalization of all aspects of our existence—from industry, through economy and war, to fashion—we must continually make whatever we do adhere to demands and needs around the globe.

The Michigan Assessment Instrument contains: specific OHSMS principles, measurement criteria for each principle, suggested measures for each measurement criterion, data collection methods, a scoring and ranking scheme, and methods for interpreting data. The goal was to construct an instrument capable of measuring the effectiveness of a wide range of occupational health and safety programs. It was felt that the construction of a single instrument would be a significant contribution . The list on Page 106 shows the 16 primary sections and 11 secondary sections of the MAI. Keep in mind that this is only a summary of headings; specific criteria summarized from the original documents detail measurement in each section.

The 15th Annual Environmental Health and Safety "White Paper," sponsored by *Industrial Safety & Hygiene News,* provides a wide range of useful "benchmarking" information for environmental, safety, and health professionals. In the 1999 issue, some alarming numbers were presented. The survey examined performance measures, and 84% identified injury, illness, and fatality data as the most important measure. The environmental, safety, and health professionals responded that benchmarked comparisons with other facilities was a low priority.[4]

BENCHMARKING AND COMPETENCIES:

Needless to say, I don't think either one of those mindsets will find wide acceptance in the future of safety performance management. Injury, illness, and fatality data will always be tracked, but its importance will begin to decrease as these more encompassing methods become familiar in the safety and health field. Benchmarking will also take on new meaning as companies realize they don't have to reinvent the wheel to address age-old safety problems.

The Michigan OHSMS Assessment Instrument Outline[1]

The MAI contains 27 sections
(16 primary, and 11 secondary)

1. Management Commitment and Resources
   - 1.1. Regulatory Compliance and System Conformance
   - 1.2. Accountability, Responsibility, and Authority
2. Employee Participation
3. Occupational Health and Safety Policy
4. Goals and Objectives
5. Performance Measures
6. System Planning and Development
   - 6.1. Baseline Evaluation and Hazard / Risk Assessment
7. HSMS Manual and Procedures
8. Training System
   - 8.1. Technical Expertise and Personnel Qualifications
9. Hazard Control System
   - 9.1 Process Design
   - 9.2 Emergency Preparedness and Response System
10. Preventative and Corrective Action System
11. Procurement and Contracting
12. Communication System
    - 12.1 Document and Record Management System
13. Evaluation System
    - 13.1 Auditing and Self-Inspection
    - 13.2 Incident Investigation and Root Cause Analysis
    - 13.3 Medical Program and Surveillance
14. Continual Improvement
15. Integration
16. Management Review

Competency-based performance management schemes are plentiful and are becoming the model for the future.[5] As it becomes increasingly important to ensure that employees have not only been trained but can demonstrate their competency, these systems will take on new meanings and increased utilization. Regulators are already stressing that training is no longer good enough; the employee must show competency.

CHANGING WORLD:

Some changes in safety performance measurement cannot help but be affected by two major phenomena:

1. Older workers will continue to comprise a larger percentage of our workforce than ever before. We will see many more employees working into their 70s and 80s than in the past. This will force the rethinking, redesign, and reorganization of many aspects of our workplaces.
2. The "virtual office." The information age is making things possible that were unimaginable just a few years ago. This, too, will affect what accommodations we must make for our employees. We will have far more employees that work out of their homes, work at a distance, or possibly that we don't ever even see or know personally. How to accommodate training for these employees and provide for their safety will provide a challenge not yet seen in the safety industry.

These far-reaching changes in the way we do business as a whole increase the need to have safety-performance measurements that focus on what is important and meaningful, not on what has always been measured before.

The writings on measurement and the research on measuring safety clearly seem to suggest the following approach:

1. Quit looking at accident-based measurements in assessing your safety system effectiveness.
2. Use audit results only when you are sure there is a correlation between the audit and safety results over time and large numbers.

3. Use a properly constructed perception survey for your primary measure and diagnostic tool.
4. Use behavior sampling on an ongoing basis for smaller units as your primary motivational measuring tool.[6]

ORGANIZATIONAL REPORTING

Safety must report directly to the top-management person. In Europe, this is required by law. Europeans realized that without this reporting structure, safety would ultimately lose out to other business sectors. If you want to see what is important in an organization, observe who reports to the top manager. Those direct reports will give you an incredibly accurate snapshot of the organization's entire culture. Without this structure, safety would forever take second place to whatever leg of the organization it reports to, whether that is human resources, operations, maintenance, or even a full "corporate safety" reporting to some manager tucked down somewhere deep in the organization.

SAFETY CULTURE

The family will become much more involved in the safety culture of our organizations. How can we expect an employee to don the appropriate safety personal protective equipment at work when the same employee goes home and mows the yard with no hearing protection, no eye protection, and no respiratory protection? This represents two drastically different cultures, two different ways of life. As long as they co-exist, we're spinning our wheels in the effort to "improve" our employees.

David Pierce describes this in his book, *Shifting Safety and Health Paradigms*, in a very visual way: "Getting dressed in the morning is not a priority—it is a value. If we arise forty-five minutes late, do we choose to get dressed when we get to work? No, getting dressed on the bus or in the elevator at work is non-negotiable. It is a societal value. We always get dressed before we leave home—even if it makes us late for work."[7]

Future metrics will most assuredly include measurement of cultural concepts such as mutual respect, trust, and credibility. These cultural fibers run deep in organizations. Without a good handle on how these fibers are doing, an organization may be spinning its wheels in whatever endeavor it may seek.

PARADIGM SHIFTS

One could not write a chapter devoted to the future of safety-performance measurement and not discuss "paradigm shifts." Paradigm shift has become one of, if not the most, widely used "buzz words" of the safety industry in the 1990s. It is interesting to read how various authors define a paradigm shift. Some convey that a paradigm shift is a large-scale cultural change or a revolutionary change in ideas and beliefs (sounds a lot like Heinrich's "Axioms" of the 1930s). Others settle for only a new model or pattern. There is no doubt that a paradigm shift is important in the long-term constant improvement of an organization, an industry, or even a country—a society. What these paradigms will be is difficult to say. Be cautious, though, when you hear these buzz words used. New programs and systems are always quick to claim the "paradigm shift victory."

EMPOWERMENT

Dr. Deming's teachings indicated that an organization should not do things just because they might influence the performance indicators; rather, we should do them because they are the right things to do. Many of the things organizations choose to debate, in terms of valid methods of improving safety performance, are merely "the right thing to do." By doing the right thing, you may never know how many injuries or accidents proactive, aggressive safety management prevented. But we must continue to do these "right things," whether or not they reduce any one specific performance indicator. Each decision to increase employee empowerment, improve workers' self-esteem, and improve interpersonal relationships between management and employees is a step toward preventing future accidents.

No one is more aware of the problems than the person who must deal with them. In almost all cases, the person who can offer some of the most ingenious solutions is the person who must deal with the problems on a day-to-day basis. Empowerment and ownership are ways to put tools in the hands of those who know just how to repair the problems.

Companies must start to see their customers as long-term partners. Organizations must seek to provide innovation, added value, and attention to the customer's needs.

REINFORCEMENT

According to Dan Petersen, "Gimmicks simply do not fit anyplace in safety theory; they make little sense in management theory and even less sense in behavior theory"[8] (posters, incentives, contests). He goes on to say, "We must recognize that to eliminate gimmicks in most circumstances is to incur the wrath of almost everyone, from the old safety director to the lowest-rated worker, for gimmicks are satisfying."[9] Systematic reinforcement of positive behavior is absolutely essential. This can be done in many different ways, just as there are many different ways to measure it. It is important now and will become even more important. Research shows that positive reinforcement is much more powerful than negative reinforcement in both behavior-building and behavior-maintaining.

NOTES AND REFERENCES

1. Weiss, Tracey B. and Hartle, Franklin, *Reengineering Performance Management Breakthroughs in Achieving Strategy through People*, St. Lucie Press, Boca Raton, FL, 1997, p. 195.
2. Redinger, Charles F., Levine, Steven P., Development and Evaluation of the Michigan Occupational Health and Safety Management System Assessment Instrument: A Universal OHSMS Performance Measurement Tool, *American Industrial Hygiene Association Journal*, August 1998, p. 572–581.
3. Redinger, Charles F., Levine, Steven P., Development and Evaluation of the Michigan Occupational Health and Safety Management System Assessment Instrument: A Universal OHSMS Performance Measurement Tool, *American Industrial Hygiene Association Journal*, August 1998, p. 572–581.

4. *Industrial Safety & Hygiene News*, January 1999, p. 21-25.
5. Weiss, Tracey B. and Hartle, Franklin, *Reengineering Performance Management Breakthroughs in Achieving Strategy through People*, St. Lucie Press, Boca Raton, FL, 1997, p. 37.
6. Petersen, Dan, *Safety by Objectives—What Gets Measured and Rewarded Gets Done*, second edition, Van Nostrand Reinhold, New York, 1978, p. 88.
7. Pierce, David, *Shifting Safety and Health Paradigms*, Government Institutes, Rockville, MD, 1996, p. 41.
8. Petersen, Dan, *Analyzing Safety System Effectiveness*, third edition, Van Nostrand Reinhold, New York, 1996, p. 124.
9. Petersen, Dan, *Analyzing Safety System Effectiveness*, third edition, Van Nostrand Reinhold, New York, 1996, p. 125.

Conclusion

There is no magic safety performance pill, likewise there is no magic safety system.

Daniel Patrick O'Brien

Success will only come with time, persistence, and investment. To succeed in the long run, it will be necessary to put up with short-term problems. A lack of determination will result in halfhearted efforts and renunciation at the first sign of difficulty. If you are committed, you will persevere and succeed; if you are not, don't bother to start.[1]

In *Jesus CEO*,[2] Laurie Beth Jones speaks of the leadership skills of one person who, with 12 people, all ordinary men reporting directly to him, made such an impact on civilization that some 2000 years later the world is still measuring time based on the date of his death. This is the kind of leadership we should strive for in our organizations. We have lost sight of excellence. If you want to lead change, then change your approach to leadership.

One thing that all safety experts will agree on—one size does not fit all. There is no magic pill. To find what works for your organization, you must diligently seek out and adapt what works for your organization. That may sound somewhat simplistic, but it is the hard reality of the safety business. Coincidentally, it is reality for the business world as well. Organization members cannot afford to spend hundreds of thousands of dollars on what was a great plan for another organization only to have it fail miserably for them.

NOTES AND REFERENCES

1. Kaydos, Will J., *Measuring, Managing, and Maximizing Performance*, Productivity Press, Cambridge, MA, 1991, p. 3.
2. Jones, Laurie Beth, *Jesus CEO: Using Ancient Wisdom for Visionary Leadership*, Hyperion, New York, 1995, p. XIII.

Index

115